POWER FOOD

James McNair

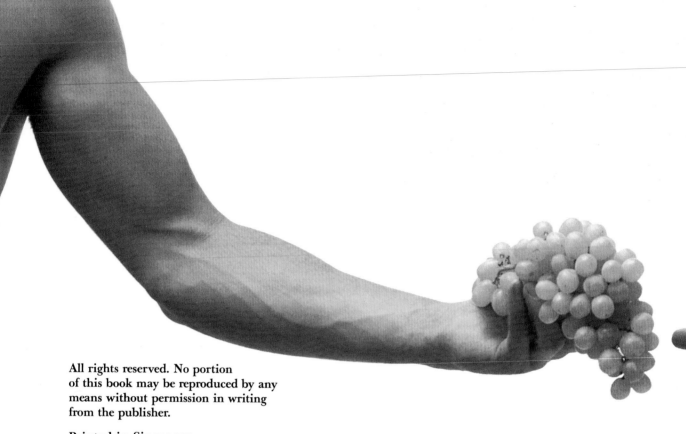

Printed in Singapore

Library of Congress Cataloging-in-
Publication Data

McNair, James K.
  Power Food.
  Bibliography: p.
  Includes index.
  1. Athletes — Nutrition.
  2. High carbohydrate diet —
    Recipes
  I. Title.
TX361.A8M39 1986   613.2'088796
  85-29938
ISBN 0-87701-369-1

Chronicle Books
One Hallidie Plaza
San Francisco, California 94102

POWER FOOD

James McNair

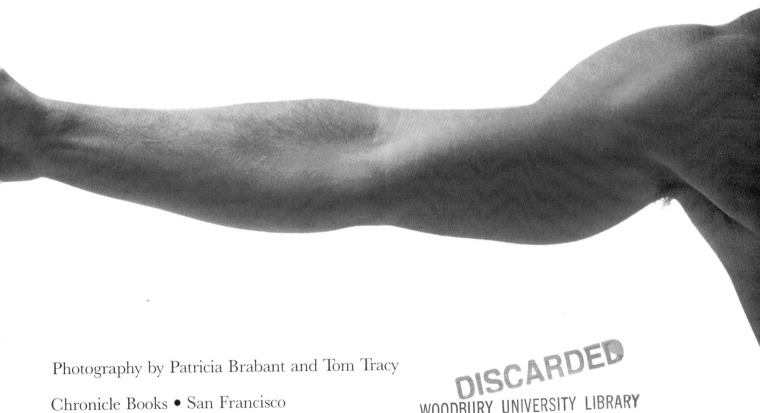

Photography by Patricia Brabant and Tom Tracy

Chronicle Books • San Francisco

Produced by The Rockpile Press, Lake Tahoe and San Francisco

Art direction, photographic and food styling, and book design
by James McNair

The Rockpile Press production by Lin Cotton
Typography and mechanical production by Terrific Graphics

Chronicle Books production by David Barich
Production assistance by Fearn Cutler
Copyediting by Rita Progler

Photography by Patricia Brabant: Front and back cover flaps, 1, 4-5,
6-7, 8, 18-19, 31, 34, 42, 46, 49, 53, 56, 61, 65, 68, 72, 76, 81, 84, 89,
92, 96, 103, 106, 109, 112, 117, 120

Photography by Tom Tracy: Front and back covers, 2-3, 10,
15, 22-23, 26-27

Assistant to Ms. Brabant: Louis Block

MODELS

David Biscardi, model: 2, 10
Jon Boone, lifeguard: back cover, 3
Brett Brown, competitive body builder:
42, 72, 112
Carolyn Chapin, sports nutritionist:
34, 49
Lin Cotton, landscape architect: 49
Donald Dragoni, physician: 6-7, 61,
back cover flap
Ira Durant, caterer: 103
Bridget Goodwin, secretary: 92
Tom Goodwin, architect: 92
Paul Harris, graphic designer: 53
Tanya High, high school freshman: 15
Rayleen Hooper, body builder:
cover flap, 84, 109
Mark Leno, sign maker: 31, 106
**Michael McKenney, fitness product
sales:** 4-5
Lucy Marcus, sixth grade student: 68
Julia Hall Parish, landscape designer,
cover
Jim Pendergast, retired welder: 96
Don Placencio, landscape foreman: 117
Stephen Sanderson, registered nurse:
1, 36, 89
Jane Stephens, fitness consultant:
1, 61, 120
Bill Strubbe, masseur: 76
**Lena Sullivan, television associate
producer:** 65
Steve Whiteman, law student: 46, 81
**Mike Zerbel, engineering graduate
student:** 8, 56

For Gene Davis, cousin and devoted friend, who inspired this book.

And with happy recollections of Heckle and Jeckle.

ACKNOWLEDGMENTS

Consultant Carolyn Chapin, Sports Nutritionist, for double checking my work.

Sue Fisher King, San Francisco, for plates, glasses, and other props. Also thanks to her very helpful staff.

The superb staff of Chronicle Books, especially Larry Smith for keeping it all running smoothly, Bill LeBlond for able editing, David Barich and Fearn Cutler for expert production, Jack Jensen and Drew Montgomery for getting us on the shelves, and Mary Ann Gilderbloom for promoting and publicizing the work.

Addie Prey, Joshua J. Chew, and Michael T. Wigglebutt for secretarial and test kitchen services, and Buster Booroo for security services.

Mike Tucker and his marvelous modem for transmitting text to type.

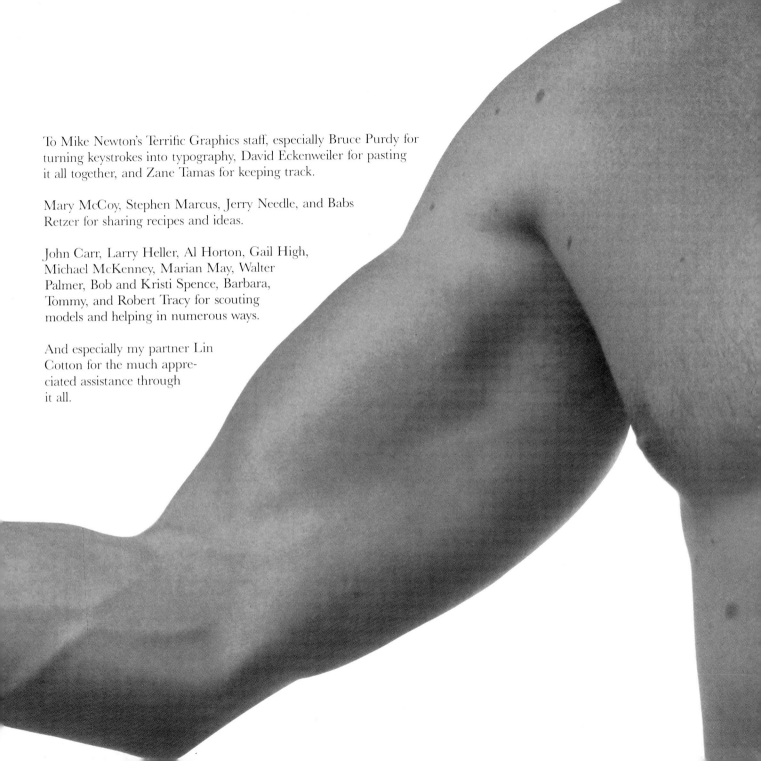

To Mike Newton's Terrific Graphics staff, especially Bruce Purdy for turning keystrokes into typography, David Eckenweiler for pasting it all together, and Zane Tamas for keeping track.

Mary McCoy, Stephen Marcus, Jerry Needle, and Babs Retzer for sharing recipes and ideas.

John Carr, Larry Heller, Al Horton, Gail High, Michael McKenney, Marian May, Walter Palmer, Bob and Kristi Spence, Barbara, Tommy, and Robert Tracy for scouting models and helping in numerous ways.

And especially my partner Lin Cotton for the much appreciated assistance through it all.

Contents

Foods for Fitness and High Performance

POWER FOOD provides active bodies with long-lasting nutritional fuel that leads to better fitness and higher performance.

Scan through all the new magazines devoted to fitness, read the piles of books on the subject, talk with professional athletes, coaches, and sports nutritionists, and you're sure to come away with a confusing array of opinions and ideas about what to eat.

Some sports nutrition pros advocate adding supplements to any diet; others say bunk to anything in the form of a pill, believing that the well-balanced diet provides all the nutrients any body needs. The question of protein source splits the group down the middle with some advocating meat for protein; others make it taboo. A few still swear by the steak-for-strength myth; most go along with the evidence supporting complex carbohydrates as the major fuel for energy. A recent study showed that extremely active people can utilize more fats than their sedentary counterparts, while much research points out a cholesterol danger from too much fat no matter how fit the body.

Rest assured that no evidence from any study to date has proven that athletes or any physically active people need anything other than a well-balanced diet that's low in fat, adequate in protein, and high in complex carbohydrates. Translated into percentages the daily diet should be comprised of a maximum of 30 percent fat, 10 to 15 percent protein, and 55 to 60 percent carbohydrate.

People who heavily exercise, compete in sports, or otherwise push their bodies to the limit can certainly indulge in more daily calories than the sedentary set. But it's the quality of what you eat that contributes to fitness and high performance. Like bodies, not all nutrients are created equally. Fats, proteins, and carbohydrates all have roles in keeping the body fit and functioning at its prime.

PUMPING IRON

In the midst of controversy over supplements or radical departures from the well-balanced diet, one thing is apparent: active people, especially runners, need additional iron.

Iron shortages result from heavy sweating and stomach bleeding due to red blood cell ruptures during strenuous exercise. Counteract by eating iron-rich foods: beef liver, red meat, poultry, and fish.

Non-animal iron sources such as spinach, legumes, and molasses are difficult to absorb, but vitamin C increases iron absorption. In lieu of iron supplements, try cooking in cast-iron ware.

Whole grain bread provides high quality performance fuel.

POLYUNSATURATED FATS

Eat at least 1 tablespoon daily of one of the following:

Fish or fish oils
Margarine, soft unsaturated types
Mayonnaise
Nuts—almonds, hazelnuts, pecans, and
    walnuts
Oils—corn, cottonseed, sesame,
    safflower, soybean, sunflower
Seeds—pumpkin, sesame, sunflower

MONOUNSATURATED FATS

Limit amounts eaten:

Avocado
Cashews
Olives and olive oil
Peanuts, peanut butter, and peanut oil

LOW FATS

Fats are used to form protective cushioning around the body's organs, insulate from cold and heat, and provide some energy for muscles. Unlike protein and carbohydrate, the body is quite capable of manufacturing its own supply of fat from unused portions of the other two nutrients. Pockets of fat are stored away for use by the body when it needs them for energy not supplied by the diet.

People concerned with achieving trim, hard bodies might think that since the body makes its own fat supply dietary fat can be entirely eliminated. The fact is that at least one tablespoon of cholesterol-lowering, polyunsaturated fats is needed daily to provide essential fatty acid or linoleic acid required by the body as catalysts for converting other foods into fats. It's also fact that no more than one tablespoon of polyunsaturated fats is needed daily by the body, and very much more can add too many fat calories and may be linked to cancer.

Saturated fats, found in animal sources and loaded with cholesterol, are not needed by the body and, except for their taste, could be totally eliminated from the diet. Monounsaturated fats found in avocado, olives, and peanuts have no effect on cholesterol in the blood and are not required by the body. By weight, all fats— polyunsaturated, monounsaturated, or saturated—contain the same amount of calories. If you need to watch your weight, fats are the first thing to cut.

Many of the foods we think we're eating for protein may actually be higher in fat than in protein. Because fats make foods taste and smell better, the average American literally "pigs out" on daily diets that are as high as 50 percent fat, while no more than 30 percent fat is recommended for good health even for highly athletic people.

Without sacrificing flavor, fats can be reduced when cooking. Use non-stick coated pots and pans. Instead of cooking in fats, coat cookware with vegetable oil sprays or wipe with vegetable oil before cooking. You can "sauté" with much less butter or other fat if you use lower heat and give constant attention to stirring. In lieu of sautéing, foods can be quickly heated in small amounts of hot stock or broth, a little soy sauce mixed with water, flavored vinegars, or a bit of tomato or carrot juice. Steam onions, celery, or carrots that will be used in dishes instead of sautéing; the results are sweet, soft, and fat free. Combine liquid-releasing mushrooms with other vegetables while cooking.

ADEQUATE PROTEIN

Simply put, protein is the building material found in every cell of the body. Protein's amino acids are used in numerous ways to form different types of structures: bone, cartilage, muscle, blood, lymph, skin, hair, nails. Nine of the twenty-two amino acids are classed as essential because they cannot be manufactured by the body and must enter through the foods we eat. Protein is not stored in the body, but must be replenished daily. Too much ingested protein is not only converted into stored fat, but causes the liver and kidneys to work overtime to process and dispose of the nitrogen component of protein.

SATURATED FATS

Restrict Amounts:

Butter
Cheese, hard or processed types
Chocolate
Coconut and coconut oil
Cream
Egg yolk
Lard
Meat, red
Milk, whole
Palm oil
Poultry, especially dark meat
Vegetable shortening

COMPLETE PROTEINS

All animal products are sources of "complete" protein, containing all of the essential amino acids. If you choose to forgo meat, bear in mind that vegetable proteins do not contain all of the essential amino acids. Certain vegetables contain some, others contain those lacking, yet are missing others. Vegetable proteins must be combined or balanced to form complete protein and provide essential amino acids.

Dried legumes combined with whole grains not only offer the most perfect nutrition possible, but they also taste great together. Throughout the centuries they've been the staples in many cuisines, as evidenced by the refried beans and rice or corn of Mexico, the Cajun and Creole red beans and rice, the Brazilian black beans with rice, the North African couscous and garbanzo beans, or the lentils and grains of India. Other ways of creating complete protein are to combine legumes with seeds or nuts, or dairy products or eggs with any source of vegetable protein.

Most nutrition authorities today are convinced that everyone needs far less protein than was thought in the past. Since protein is not used to fuel muscles unless all carbohydrate and fat stores are depleted, there is no reason to believe that very physically active people need or use more protein than those involved in normal activities. Exceptions to the rule may include still-developing adolescent athletes, people recovering from injuries who need extra protein for rebuilding tissue, and body builders striving for adding muscle bulk; even this group will receive adequate protein by eating no more than the U.S. Recommended Dietary Allowance (RDA). This system, based upon age and size, set up by The Food and Nutrition Board of the National Academy of Sciences has a built-in extra 45 percent safety margin. This means, according to the majority of nutritionists, that no more than two-thirds of the recommended amount of protein will suffice for most of us, even sports professionals.

There is certainly no need for added protein powders so popular with athletes and body builders. Instead of enriching your muscles you're simply lining the pockets of manufacturers.

Whole grains on left of gymnast can combine with legumes on right to form complete vegetable protein.

CARBOHYDRATE LOADING

Some endurance athletes report favorable results from storing up glycogen through the practice of carbohydrate-loading, usually a three-day program preceding the big event in which carbohydrates make up 70 percent of the daily diet. There is no evidence that carbohydrate-loading is helpful or healthy for activities that are strenuous but short and even endurance athletes are advised not to practice carbohydrate-loading except occasionally in preparation for major events.

HIGH COMPLEX CARBOHYDRATES

Carbohydrates provide powerful fuel for energy to keep the body fit when stored as glycogen in the muscle tissue and liver, waiting to be broken down into glucose or blood sugar as needed. Fortunately for those of us who love eating, the body can only store enough glycogen for a few hours at a time; active people need to replenish fuel by eating carbohydrates several times a day.

In addition to glucose made from glycogen, free fatty acids are also utilized as fuel during light to moderate exercise. As activities become more strenuous or enduring, muscles rely on their own crucial supply of stored glycogen. Exhaustion or burn-out occurs when this supply runs out. From this it is apparent how important it is that muscles have a steady supply of good glycogen. Studies dealing with various mixes or ratios of foods have repeatedly demonstrated that muscles run better and longer on the chains of energy created by carbohydrates.

Carbohydrates come in two forms: simple and complex. Simple carbohydrates include foods made with refined and processed sugars, while complex forms or starches include whole grains, dried legumes, and fresh vegetables.

While it's true that a calorie is a calorie when it comes to adding weight, a calorie is not a calorie when it comes to powering the body. Complex carbohydrates offer long-lasting energy; the same calories in simple carbohydrates or sugars are "empty," failing to keep the body going. Fats cause foods to taste great, but they add tremendously to the calorie count. By eating fat-free starches or complex carbohydrates with minimal fat added to enhance flavors, you could enjoy a far more filling and satisfying meal, using the same number of calories to provide better performance fuel.

NUTRITIONAL FUEL

Since the Senate Select Committee on Nutrition and Human Needs released its findings in 1977, fitness-conscious Americans have led the way in following the Committee's recommendations: more whole grains, legumes, fresh vegetables, fruits, fiber; less fats, cholesterol, salt, refined and processed sugars, and refined flours.

Well-balanced nutrition can be accomplished through scrutinized vegetarian diets, with or without dairy products. The surest way to optimal nutrition includes a wide variety of foods, insuring that all bases are covered.

Listen to your body to determine what and how much to eat. Food cravings may be calling for better nutrition, not specific foods. When you want sugar, eat carbohydrate snacks that provide long-range energy, not instant gratification that quickly lets you down. Experiment to see how certain foods affect you, what makes you run better or endure longer workouts. Above all enjoy a wide variety of all kinds of foods for ultimate nutrition.

Forget rigid diets, and practice a winning way of eating that can easily satisfy you for the rest of your life. For the super active and even the moderately active, high complex carbohydrates provide the fuel for fitness and higher performance. Let's look at the major carbohydrate sources.

U.S. DIETARY GOALS

Established by the Senate Select Committee on Nutrition and Human Needs chaired by George McGovern.

58% Carbohydrates
    48% complex carbohydrates and naturally occurring sugars
    10% refined and processed sugars
12% Proteins
30% Fats
    10% polyunsaturated
    10% monounsaturated
    10% saturated

GUIDELINES FOR MEAT EATERS

Choose lean chicken, turkey, and fish far more frequently than red meats. For those times when you do go the red meat route, select lean cuts instead of the once-favored marbleized cuts.

No matter what the meat, practice simple cookery—grilling, broiling, poaching, or roasting—with very little or no added fat. Keep meat servings small; about six to eight ounces per day is sufficient and healthier. Fill up with low-fat carbohydrates such as pasta, grains, and high-fiber vegetables.

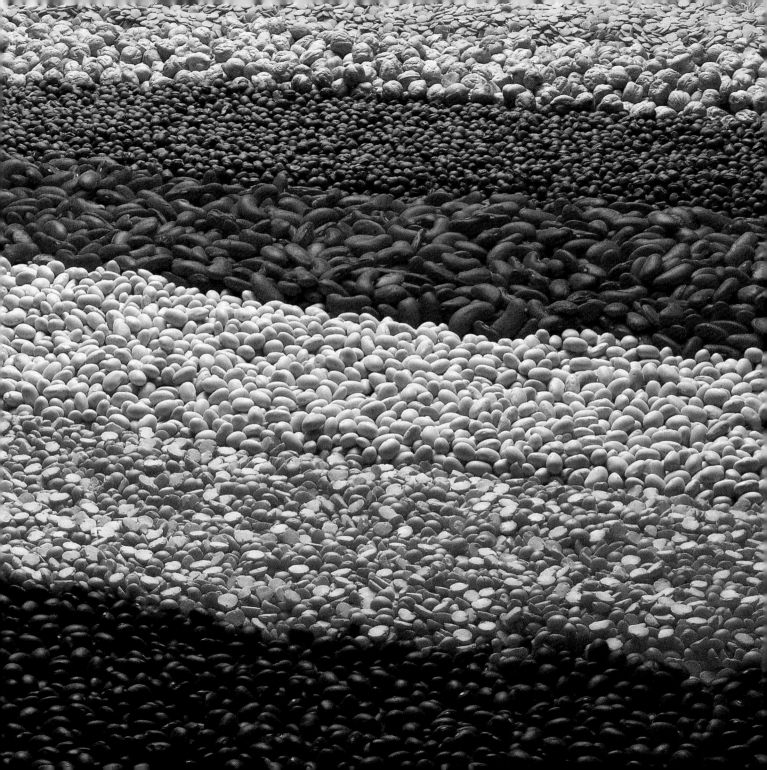

STOCKING UP ON WHOLE GRAINS

Store grains in tightly sealed glass jars in a cool dry place or refrigerate during hot weather to keep germ oils from turning rancid. Stored this way, they last indefinitely. Whole forms can be ground into desired form of meal or flour as needed or purchased in a number of pre-ground or cut forms.

- ☐ Amaranth
- ☐ Barley
- ☐ Buckwheat
- ☐ Bulgur
- ☐ Corn
- ☐ Millet
- ☐ Oats
- ☐ Rice, brown
- ☐ Rye
- ☐ Sorghum or milo
- ☐ Triticale
- ☐ Whole wheat
- ☐ Wild rice

GRAINS

Whole grains, so underused in the American kitchen, are one of the world's oldest forms of power food. Referred to as the "fuel of civilization," grains in some form have been the "staff of life" throughout history: the rice of the Orient, the sorghum and millet of central Africa, the couscous of northern Africa, the buckwheat of eastern Europe, the oats of the British isles, the maize or corn of the Indian Americas. Early American settlers brought with them the grains of their homelands that flourished, yet, by some bizarre twist, modern Americans have come to favor refined white flour and highly refined versions of other grains over tastier and far more nutritious whole grains.

No food source offers more to the fitness-conscious meal planner than whole grains. Experiment with cooking and using varied grains in favorite family recipes. Basic recipes on pages 38-41 will get you started. Once cooked, grains may be eaten hot or cold as cereals or added to other dishes for satisfying nutrition.

Like fellow complex carbohydrate vegetables and fruits, whole grains have a lot of indigestible fiber that passes through the large intestine virtually intact. In the colon, some of these fibers are fermented and absorbed into the body as volatile fatty acids. A fiber-rich, starchy carbohydrate diet helps control weight because necessary long chewing satisfies the appetite with less food and fibrous bulk actually fills you up sooner. Fiber also helps move all food through the digestive tract faster, disposing of some of the fat and protein consumed before they are absorbed into the body.

Overleaf: Grains on left and legumes on right represent a huge array of powerful potentials.

LEGUMES

Hanging pods growing on vines or bushes of beans or peas have a great power food stored inside. Nutritious legumes are available fresh in season or dried at any time.

Led by soybeans, legumes contain the highest percentage of protein of any vegetable. Many compete with meats in quantity and quality of protein when deliciously teamed with grains or seeds.

Tofu is a soft curd or cheese made from soybeans. Low in saturated fats and calories, high in protein, vitamins, and minerals, tofu is easily digested and extremely versatile. Associated with Asian dishes where it's been used since civilization began, tofu is a good protein source that can be added to Western dishes. When first frozen, then thawed, drained, and squeezed of excess moisture, it takes on the texture of ground meat. When cooking with tofu, keep in mind that lots of spices and flavorings can be sponged up by the bland curd; generous seasoning and lengthy marinating can add much flavor. And remember to team it up with seeds, nuts, or grains for essential amino acids.

FRESH SEASONAL LEGUMES

- [ ] Fava beans
- [ ] Field or crowder peas
- [ ] Green, string, snap or wax beans
- [ ] Green, sweet, or spring peas
- [ ] Lima or butter beans
- [ ] Snow, sugar, sugar-snap, or edible pod peas

DRIED LEGUMES

Store in airtight, glass jars:

- [ ] Black-eyed or cow peas
- [ ] Black or turtle beans
- [ ] Cranberry or pink beans
- [ ] Garbanzo beans, also known as chick-peas or ceci-peas
- [ ] Green flageolet beans
- [ ] Green peas, whole or split
- [ ] Haricot beans
- [ ] Lima beans or baby limas
- [ ] Pinto or chili beans
- [ ] Kidney or red beans
- [ ] Soybeans
- [ ] White, navy, Great Northern, marrow, cannellini, or white kidney beans
- [ ] Lentils, green, orange, or brown
- [ ] Peanuts
- [ ] Pigeon peas
- [ ] Yellow peas, whole or split

FRESH VEGETABLES

In the vastly differing world between artichokes and zucchini are great sources of carbohydrates, vitamins, minerals, trace elements, natural sugars, and fiber, all totally without cholesterol. Vegetables offer satisfying, low-calorie volume to the diet.

For optimal nutritional benefits, eat vegetables raw. Pick them off the vine or buy the freshest vegetables possible daily to avoid vitamin-reducing refrigerator storage. Although not as tasty as garden fresh produce, frozen vegetables offer better nutrition than wilting or long-stored fresh versions. In many cases frozen vegetables are vine-ripened and quickly processed, preserving more nutrients than fresh produce that might take days or weeks in shipping and storage before reaching your table. The nutrients in canned vegetables have been grossly reduced through heat processes. The only canned vegetable I would ever think of using is Italian-style tomatoes, which often have more flavor than fresh supermarket varieties.

When you cook vegetables, be careful to avoid destroying nutrients from overcooking and don't throw valuable nutrients away with the boiling water. Steaming, stir-frying, baking, sautéing, or microwaving not only assures better-tasting results, but retains more of their considerable nutritional value.

Overleaf: Fresh vegetables offer variety in taste and nutrients.

FRUITS

Lusciously ripe fruits nutritiously satisfy sweet cravings due to their naturally occurring sugars ranked among the good-guy carbohydrates. Seasonal fruits are perfect for snacking or almost instant low-calorie desserts.

Like vegetables, fresh fruits are rich sources of vitamins, minerals, trace elements, and fiber, void of fat or cholesterol. In fact, the pectin, a fiber found in several fruits including apples, even helps lower cholesterol levels.

As with vegetables, fresh and raw is the best way to consume all the nutritional benefits offered by fruits. For a change of pace, stewed, poached, or baked fruits make tasty and healthy breakfast additions or anytime sweets.

If you're watching your weight, as most athletes and fitness-conscious people usually are, keep in mind that dried fruits contain highly concentrated sugars and should be enjoyed in moderation. Still they're better than refined sugars for adding sweetness to baked goods and other foods.

CHOCOLATE FOR QUICK ENERGY?

Fiction: Chocolate provides instant energy for power.

Facts: The feeling of quick energy after eating chocolate results from a rise of glucose (blood sugar) and is not good for athletes, be they pros or weekenders. Such quick forms of glucose trigger the liver to halt normal glucose production. When the temporary sugar supply is used up, no replacement glucose is readily available, leaving one jittery and uncoordinated, with feelings of confusion and weakness.

Even if the aftereffects of a quick rush were worth it, chocolate isn't digested quickly enough for activities where quick bursts of energy are needed. During endurance activities, not only is there always the danger of running out of glucose, but the 50 percent of saturated fat calories in chocolate is not easily absorbed and thus results in cramping and slowing down.

As quick power food, fruits are far superior to chocolates.

Overleaf: Enjoy naturally forming sugars of fresh fruits for dessert or snacks.

LOW-FAT OR
NONFAT DAIRY PRODUCTS

Cheeses — low-fat or uncreamed cottage,
farmer, pot, part-skim ricotta,
part-skim mozzarella, Parmesan,
low-fat cheese slices

Ice milk or sherbet

Milk — nonfat or skimmed, low-fat,
evaporated low-fat, low-fat
buttermilk

Yogurt, nonfat or low-fat

CUTTING BUTTER FAT

For cooking or spreading, a little butter
can go a long way towards enhancing
flavors. Use it judiciously and enjoy.
You might purchase blended butter-
margarine products or make your own
by combining equal parts unsalted butter and unsalted margarine or vegetable
oil. True, calories are the same, but
saturated fats will be lower and taste
will be much better than plain
margarine.

DAIRY PRODUCTS

Fats supply most of the calories in dairy products made from whole
milk. Milk products, with the butterfats wholly or partially skimmed
away, are good food sources of vitamins, minerals, protein, and simple
carbohydrates. Remember to keep simple carbohydrates restricted to no
more than 10 percent of your total daily diet.

Low-fat milk can replace cream in cream soups, sauces, frozen desserts,
and baking. Yogurt and buttermilk, made from skimmed or partially
skimmed milk, are excellent products to add smoothness and richness
to cooking.

Those seeking fit bodies should watch out for hard or semi-soft cheeses
whose calories contain 80 percent saturated fat. Enjoy small servings of
rich cheeses only occasionally. The rest of the time, stick with small portions of Parmesan, Romano, and mozzarella made from partially
skimmed milk with fat reduced to around 20 percent. Or try sliced
low-fat cheeses made with unsaturated oils replacing part of the butterfat; however, be sure to check labels, making sure fat-rich coconut or
palm oils were not used to replace the butterfat.

Soft cheeses made from wholly or partially skimmed milk, such as cottage cheese, farmer cheese, pot cheese, and ricotta, are good protein
alternatives to meat. Eliminate or greatly restrict use of creamy cheeses
such as brie, triple creams, or even cream cheese. It's true, as the ads
say, that cream cheese contains less fat than butter or margarine, but it's
still high in fat and cholesterol. Smooth Yogurt Cheese (page 58) is so
good you won't even miss cream cheese.

EGGS

Eggs are an inexpensive source of fairly complete protein and are used extensively in baking with complex-carbohydrate-rich grains or combined with vegetables to complete the protein. Like other good things, too much can be harmful. To maintain the low-fat, low-cholesterol way of eating, consume no more than three to four cholesterol-rich yolks per week, but feel free to enjoy an unlimited supply of egg whites. Two whites successfully substitute for each whole egg in most recipes, except souffles or egg-rich cakes.

Limit intake of noodles made with eggs. Read labels; eggs in ready-made foods count too.

SALT

Whether from dietary salt or salt tablets favored to counteract sweating, too much sodium in the fluids outside of cells drains water from cells and body fluids to dilute the excess salt. Cells, including muscle cells, are left dehydrated, leading to poor performance. Water drained from the blood supply to dilute excess salt affects the cooling process of sweating and can lead to heat prostration, slows down transport of nutrients, and interferes with body waste elimination.

Instead of replacing salt lost from sweating with more salt, drink plenty of water to counteract the draining of cellular and body fluid salts. If a workout sweats out as much as ten pounds, sprinkling a little salt on food quickly puts the system in balance again.

Like everyone else, very physical people function best when salt intake is restricted. Low-salt diets condition the body to conserve salt, reducing sweating and leading to better performance. In lieu of salting, sprinkle foods with fresh herbs, spices, or lemon juice for added flavor.

ALCOHOLIC DRINKS

Unlike other carbohydrates, beer, wine, and other alcoholic beverages cannot be used as direct sources of energy fuel for muscles. They should be avoided before or during workouts or competitions, not only because they lessen coordination, but their diuretic effect can cause dehydration.

According to sports nutritionists, after heavy physical activity moderate portions of alcohol may be acceptable. However, they should not be used to replenish lost fluids or electrolytes, for they only dehydrate. Fill yourself with water before indulging in an alcoholic drink.

BEVERAGES

Water should be dubbed the drink of champions. Competing athletes as well as the daily jogging set can hurt both the performance and the body if insufficient water is consumed. Water is essential to keep bodily functions running at their peak; without enough water, muscles become dehydrated, leading to fatigue. During activity, plain water wins hands down over carbonated beverages, which can cause bloating gas.

Develop the habit of drinking a lot of water before, between, and during meals. The standard ratio calls for one quart of water daily for every one-thousand calories eaten. For hungry competitors and those in heavy training, this can mean a lot of water.

Heavily sugared drinks and fruit juices with a high concentration of sugars slow down absorption of fluid from the stomach and actually increase the need for water. If you choose to drink something other than water during competition or training sessions to prevent heat stress, choose those with a concentration of less than 2.5 percent sugar. If you enjoy fruit juice, mix with water to reduce concentration. During endurance competition, when you need quick replenishment of carbohydrates, concentrated sugar in beverages can be increased to 5 percent. Drinks containing glucose polymers now on the market may be beneficial to some.

Most current nutritional thinking says that caffeine beverages in moderation are okay for healthy people. Most sports experts recommend that coffee or tea be avoided just before or during competitions or workouts. Even though caffeine can be a perk-up, resulting in slightly longer endurance or lessening muscle fatigue, it can also overstimulate, cause gastrointestinal activity, or work as a dehydrating diuretic and flush out valuable body water and salts. Of course, coffee or tea, hot or iced, should be without fat-rich cream or empty calorie sugar; low-fat milk is great.

Water: the champagne of champions!

HOW AND WHEN TO EAT

Good eating habits make you more energetic, leaving you trimmer and more resistant to disease. Maintaining a regular mealtime schedule, without skipping meals, and spreading out daily intake of calories fairly evenly, conditions the body to not store up food as fat and gives you consistent energy throughout the day. The optimal eating plan would allow time for four to six small meals each day.

Enjoy the art of dining, not just eating. Many active people stay on the run, grabbing food just to keep them going. Nutritionally and emotionally, your body will benefit from the exercise of gracious dining as much as it does from the hard work you put into it. Even if you're dining alone, set an attractive table, create an atmosphere of relaxation, and sit down to leisurely indulge yourself. Chewing food slowly starts the digestive processes and stimulates a satisfied feeling with smaller amounts of food; you actually eat less when you give the body time to tell you when it's had enough.

Realistically speaking, people on the go can't always take time to prepare and eat nutritious, leisurely meals at home. When you must eat out, the best restaurants for low fats, adequate protein, and complex carbohydrates include the Oriental, Mexican, and Italian. When dining on all-American fare, select simply cooked chicken or fish, vegetables, baked potatoes (sans sour cream and butter), and fresh fruit for dessert. Don't hesitate to request that butter or sauces be withheld. Ask for fresh lemon or oil and vinegar to drizzle over salads, and low-fat milk for coffee instead of cream.

When you slow down for fast foods, go for Mexican bean, rice, and corn tortilla dishes, or nearly perfect pizza with vegetable toppings. Work in a visit to salad bars for fiber.

Eat a high carbohydrate meal preceded and accompanied by plenty of water about three hours before athletic competition or heavy workout. Avoid fibrous foods that can cause gas such as raw vegetables, most raw fruits (bananas are a great exception), nuts, beans, and whole grains, as well as proteins and fats that require a long digestive period.

During endurance events, easily digested fruits such as bananas offer quick restoration of carbohydrates. Plenty of water is essential throughout and following any strenuous activity.

Just as muscles of the body become stronger through participation in competitive team sports, individual weight lifting, group aerobics, gymnastic workouts, bicycling, running, walking, skiing, or any other exercise you choose, so does the body condition and strengthen itself through nutritious eating. Food should be thought of as much more than fuel for powering muscles. Wise choices and good eating habits also provide underlying health and stamina for the most rigorous activities. Food that is good for you can also be food that tastes good, satisfying one of life's greatest pleasures.

FURTHER READING

Bailey, Covert. *Fit or Fat?* Boston: Houghton Mifflin Company, 1977.

Brody, Jane. *Jane Brody's Nutrition Book.* New York: W. W. Norton & Company, Inc, 1981; Bantam Books, 1982.

Darden, Ellington. *The Nautilus Nutrition Book.* Chicago: Contemporary Books, Inc., 1981.

Pritikin, Nathan. *Diet for Runners.* New York: Simon and Schuster, Inc., 1985.

Smith, Nathan J., MD. *Food for Sport.* Palo Alto, California: Bull Publishing Company, 1976.

Recipes

A very practical food editor once told me that a cookbook is worth its total cost and effort if only one good recipe comes out of it. I hope that among the following pages you find several worthy additions to your cooking repertoire.

Obviously, a volume this size cannot begin to be a definitive work on the vast subject of nutritional cookery for fitness. My collection of recipes represents a sampling of the dishes active people should eat, emphasizing minimal fat, adequate protein, and optimal complex carbohydrate.

Although I personally disdain either "gourmet" or "health" labels on foods that taste good or are good for you, in the accepted usage both those terms might be applicable to these recipes. All are in keeping with the new dictates of American cuisine—a wide variety of the finest and freshest ingredients available cooked simply. No longer does healthy food have to be reminiscent of stereotyped "health" food, tasting bland and looking like dinner for the dogs. Experiment with the whole world of seasonings and you'll never miss salt, cream, or butter. Keep in mind that the appearance of food and style of presentation is as important to its ultimate enjoyment as great looks are to the physically fit. Arrange food beautifully, adding finishing touches that make these naturally healthy dishes appealing to the discerning eye as well as the palate.

I've tried to be realistic in my approach to eating for power and fitness, creating recipes that will become old friends. Very few people, for example, will totally give up sweeteners, thus you'll find a little sugar, honey, or maple syrup suggested here and there. I may suggest a bit more oil than Bailey, Haas, or Pritikin advocate, but still keep these within recognized nutritional limits.

Cooking good food can be as much fun as an
exhilerating workout.

OTHER COOKBOOKS

Brody, Jane. *Jane Brody's Good Food Book*. New York: W.W. Norton & Company, 1985.

Claiborne, Craig with Franey, Pierre. *Craig Claiborne's Gourmet Diet*. New York: Times Books, 1980.

Cunningham, Marion. *The Fannie Farmer Baking Book*. New York: Alfred Knopf, Inc., 1984.

Hewitt, Jean. *The New York Times Natural Foods Cookbook*. New York: Quadrangle Books, 1971.

Hurd, Frank J., Dr. and Mrs. *Ten Talents*. Chisholm, Minnesota: Hurd, 1968.

Lappe, Frances Moore. *Diet for a Small Planet*. New York: Ballantine Books, Inc., 1971, revised 1984.

London, Sheryl and Mel. *Creative Cooking with Grains & Pasta*. Emmaus, Pennsylvania: Rodale Press, Inc., 1982.

With today's emphasis on reduced sodium, you'll find no salt in my recipes, except in some baked goods and other dishes where it would be inappropriate for salt to be added at the table by those who indulge; however, even in those recipes salt is listed as optional. If you use salt in moderation without consequences and know those you're cooking for also enjoy it, then by all means add it to taste along with other seasonings.

Use my suggested quantity of servings only as a guideline since portioning varies greatly according to when the food is eaten, number of calories burned during physical activity, and appetite.

You'll find no calorie count or nutritional component analysis accompanying my recipes. Relax and enjoy good eating as an art, not a science; all ingredients and cooking methods in this collection adhere to the principles outlined earlier so you can enjoy eating without guilt or worrying about minute degrees of variance from rigid rules.

Many of the recipes may be enjoyed at any time of the day. For example, crunchy Blue Corn Pancakes are as good at midday or evening meals as they are in the morning. Pizza is included with other breads, but makes a perfect main course. Maple Custard is appropriate at the fanciest dinner party, a family breakfast, or as an afternoon snack. Quite a few recipes are interchangeable courses. Small bowls of Cajun Seafood Gumbo, for example, can be a first or second course; larger bowls can become the only course.

BASICS (38-43). Start by learning to cook whole grains and dried legumes used in so many of the recipes that follow. Once you've mastered these two groups, create your own new combination dishes.

BREADS AND BREAKFASTS (44-53). Pancakes, waffles, muffins, and quick breads are the bases of energy-boosting breakfasts, but they make great anytime snacks or suppers as well. Pizza, my favorite bread, ends this section.

STARTERS (54-59). Here are a few light ideas to spark appetites.

SOUPS (60-63). Small servings being a meal; larger bowls can become a satisfying meal. Also see the gumbo recipe on page 83.

SALADS (64-74). More substantial than the mixed green counter-parts, these salads are based on grains, legumes, and fish.

PASTAS (75-82). Serve pasta in small portions as first courses or larger portions as the main events most Americans prefer.

MAIN EVENTS (83-99). Whether starring fish, poultry, meat, or vegetables, here are both down-home good and fancy dishes for lunch or dinner. Some vegetable dishes can also be sideliners.

SIDE DISHES (100-109). Several of these starch or vegetable dishes are hearty enough to be made into high carbohydrate whole meals with the addition of soup or salad.

SWEETS (110-117). When plain fresh fruit just isn't enough, try one of these not-too-sweet yet satisfying desserts to end a meal or as a pick-me-up snack.

Cooked Whole Grains

Cooked grains can be eaten immediately as a breakfast cereal or anytime snack; add milk, fruit, and other favorite toppings. They may be served also with a savory sauce or used in stir-fry dishes, stuffings, puddings, casseroles, and other dishes. Most can be successfully refrigerated in covered containers for up to one week. Zap briefly in microwave or reheat with a little liquid in a small pan over low heat.

Raw whole grains should be picked over to dispose of any foreign matter, then rinsed in cold water. Do not rinse cornmeal or rolled oats. For added flavor, grains that will be used for pilafs and other savory dishes can be cooked in vegetable, poultry, or meat stock instead of water.

Directions that follow are for cooking the most readily available whole grains. Many also come in flake or grit form; cook according to directions on manufacturer's packages.

1   cup amaranth

3   cups cold water

AMARANTH: Place amaranth and water in a saucepan over medium-high heat, bring to a boil, cover, reduce heat to very low, and simmer until liquid is absorbed and grains are tender but chewey, about 25 minutes. Makes 2½ cups.

BARLEY: Bring water to a rolling boil, slowly stir in barley, return to a boil, cover, reduce heat to very low, and simmer until liquid is absorbed, about 45 minutes. Remove from heat, uncover briefly and fluff with a fork, cover pot with a layer of paper toweling, replace cover, and let stand 10 minutes. Makes 4 cups.

4	cups water or stock
1	cups hulled or pearled barley

BUCKWHEAT (KASHA): Combine egg with buckwheat in a wide-bottom pot over medium heat. Stirring constantly, cook until grains are separated and coated with the egg, about 2 to 3 minutes. Add water or stock along with oil and stir. Cover tightly, reduce heat to very low, and simmer until liquid is absorbed, about 5 minutes for fine kasha, 8 minutes for medium kasha, and 10 to 15 minutes for whole groats. Makes 2 cups.

1	egg, lightly beaten
1	cup unroasted or roasted buckwheat groats
2	cups boiling water or stock
1	teaspoon vegetable oil

BULGUR (PRECOOKED WHEAT): Place bulgur in a bowl, add water or stock, cover and let soak; 1 hour for cracked, 2 hours for whole. Drain in a sieve lined with cheesecloth, making a bag out of the cloth and squeezing out excess moisture. Makes 3 cups.

1	cup whole or cracked bulgur
3	cups boiling water or stock

CORNMEAL: In a small bowl, combine cornmeal with 1 cup cold water. Slowly pour cornmeal mixture into 3 cups boiling water in a saucepan, stirring constantly with a wire whisk until water returns to a boil. Cover pot, reduce heat to very low, and simmer for 5 to 10 minutes. Makes about 3 cups.

1	cup cornmeal
4	cups water

3   cups water or stock

1   cup polenta

CORNMEAL, POLENTA: Bring water or stock to a boil in a heavy saucepan or copper polenta pan. Gradually add polenta in a steady stream, stirring constantly with a wooden spoon. Continue to stir frequently until thickened to your preference, 25 to 35 minutes. For sliceable polenta, the grain should come away from the sides of the pot and be able to support the wooden spoon upright. Add cheese or butter if desired and pour onto a platter or wooden board and smooth surface with a water dampened spoon or spatula. Thin polenta should have a texture similar to cream of wheat and can be poured directly into serving bowls. Makes 3 cups.

---

1   cup millet

2   teaspoons vegetable oil

2   cups boiling water

MILLET: For fluffy millet with separated grains, place millet and 1 teaspoon oil in a skillet over medium heat and cook until lightly browned and nutty smelling, stirring or shaking continuously. Pour toasted millet into a saucepan containing water, add remaining 1 teaspoon of oil, stir once, cover, reduce heat to very low, and simmer with cover intact until done, about 25 minutes. Remove from heat, remove lid, cover top of pot with paper toweling, replace lid, and let stand 10 minutes. Fluff with fork. Makes 4 cups.

---

1   cup whole oat groats or rolled oats

     Water

OATS: Slowly pour groats into 3 cups boiling water in a saucepan. Stir once and cover tightly. Remove from heat and let stand for about 25 minutes. To cook rolled oats, slowly pour into 1½ cups boiling water in a saucepan. Stir once and cover tightly. Remove from heat and let stand for 8 to 10 minutes. For creamier rolled oats, combine oats with 1 cup cold water and 1 cup milk. Bring to a boil, reduce heat to very low, and simmer, stirring frequently, for 10 minutes. Makes 2 cups.

RICE, BROWN: Bring water or stock to a boil in a large, wide-bottom pot; add oil. Slowly stir in rice and return pot to a boil. Reduce heat to very low, cover tightly, and simmer for 45 minutes. Remove from heat, fluff with a fork, cover again, and let stand for 10 minutes. Makes 4 cups.

2½	cups water or stock
1	tablespoon vegetable oil
1	cup long- or short-grained brown rice

RICE, WILD: Bring water or stock to a boil in a wide-bottomed saucepan. Stir in rice and return to a boil. Reduce heat to very low, cover, and simmer until grains pop open, about 45 minutes. Remove from heat, drain off any excess water, cover pot with paper toweling, replace lid, and let stand until dry, about 5 minutes. Fluff with a fork. Makes 4 cups.

4	cups water or stock
1	cup wild rice

RYE: Soak rye overnight in 2 cups cold water. Drain, reserve and measure soaking water, adding enough cold water to make 2½ cups. Bring water to a boil and add rye. Return to a boil, stir, reduce heat to very low, cover tightly, and simmer until liquid is absorbed, about 45 minutes. Remove from heat, cover pot with paper toweling, replace cover, and let stand for about 10 minutes. Makes 2¼ cups.

1	cup whole rye berries or groats
	Water

WHEAT: Place wheat berries and oil in a skillet over medium heat and cook until lightly browned and nutty smelling, stirring or shaking continuously. Cover toasted wheat with water and soak overnight. Drain, reserve and measure liquid, add enough cold water to equal 3½ cups. Bring water to boil in saucepan, pour in wheat, return to a boil, cover, reduce heat to very low, and simmer until wheat tests done when you bite into a kernel, about 1 hour for whole berries, plus 10 extra minutes for hulled berries. Cover pot with paper toweling, replace lid, and let stand for 10 minutes. Makes 2¾ cups.

1	cup whole or hulled wheat berries
1	teaspoon vegetable oil
	Water

Cooked Dried Legumes

Cooked beans may be seasoned as desired and served alone or with grains. They may be added to soups, stews, and casseroles.

Rinse beans several times under running cold water, picking off and discarding whatever floats to surface.

With the exception of black-eyed peas, lentils, and split peas, dried beans require soaking. To soak, place beans in a mixing bowl and add enough water to cover beans by several inches. Soak for several hours or overnight. Large beans such as fava and white beans may require long soakings. Soaking soybeans can easily ferment during hot weather; refrigerate during their soak to be on the safe side.

If you forget to presoak beans, all is not lost. Cover with water in a large pot, bring to a boil, turn off heat and let stand for 1 hour before proceeding with cooking.

Discarding the soaking liquid before cooking gets rid of some of the substances that cause gas in some people, but also gets rid of some of the nutrients and flavor. A pinch of baking soda added to the cooking water helps reduce problematic gas. The most nutritious way to cook beans is to simply pour the beans along with their soaking water into a heavy pot, adding more water if needed to cover well, being sure the pot is not more than three-quarters filled. Rub a little oil around the top edge of the pot to help keep water from boiling over. Never add salt, fresh tomatoes, or tomato products until beans are tender; all inhibit softening of skins. Bring to a boil over high heat, cover partially, reduce heat to low, and simmer until beans are soft, adding water as required; about 30 minutes for lentils, blackeyed or split peas, 1 to 2 hours for others. High altitude, too little soaking, or hard water all add to length of cooking time.

Pressure cookers save time and hold in nutrients. Follow manufacturer's directions. If you overcook a batch or two until you get the hang of your cooker, enjoy some bean soup.

Makes 4 to 6 cups

1	pound (about 2 ¼ cups) dried beans
7-8	cups water
	Vegetable oil

It's hard to beat the combined taste and nutrition of Creole Red Beans and Rice, recipe on page 99.

Fruity Cottage Cheese Pancakes

6	eggs, separated
2	cups cottage cheese
¼	cup oil
2	tablespoons honey
1	teaspoon vanilla
¼	teaspoon salt (optional)
½	teaspoon cinnamon
1	teaspoon grated lemon peel
1	cup unbleached all-purpose flour
1	cup sliced bananas or fresh berries

Cloud light and yummy! Cottage cheese acts much like tofu as a vehicle for other flavors, in this case fresh fruit.

Blend egg yolks with cottage cheese, oil, honey, vanilla, salt, cinnamon, lemon peel, and flour in a medium-sized mixing bowl. In a separate bowl, beat egg whites until they form light peaks, but are not dry. Fold into mixture. Gently stir in bananas, if desired.

Spoon onto lightly oiled preheated griddle and bake until tops are bubbly. If using berries, sprinkle over batter as soon as it is spooned onto griddle. Turn and cook until bottoms are done. Serve hot with selected toppings.

Serves 4 to 6.

Low-fat Granola

5	cups rolled oats
¼	cup whole wheat flour
¼	cup nonfat dry milk
¾	cup sesame seeds
1	cup sunflower seeds
1	cup chopped almonds
2	teaspoons grated orange rind
½	cup frozen unsweetened apple juice concentrate, thawed
1	cup hot water
1	cup chopped dates
1	cup raisins

Most granolas are coated with oil or butter and heavily sweetened with honey. Try this version for a reduced-calorie, high-energy breakfast or snack. Store in tightly sealed plastic bags or jars in a cool place.

In a large mixing bowl, combine oats, flour, dry milk, sesame and sunflower seeds, almonds, and orange rind. Mix well. Add apple juice concentrate and water and mix thoroughly.

Thinly spread mixture in shallow roasting pans and bake at 300°F until dry and toasted, stirring occasionally, about 40 to 45 minutes. Pour into large mixing bowl and cool slightly before stirring in dates and raisins. When completely cooled, store in tightly sealed glass jars.

Makes about 10 cups.

Grain-Rich Berry
Waffles or Pancakes

Don't be fooled by the ingredients list into thinking these will be heavy; they turn out light and scrumptious. When berries aren't in season, they're delicious plain or with sliced bananas.

Combine oats, buckwheat, whole wheat flour, cornmeal, baking powder, and baking soda in a large mixing bowl.

In a separate large bowl, combine buttermilk, yogurt, syrup, eggs or egg whites, and oil and beat until well blended. Add dry ingredients and blend together quickly. Allow to stand for about 15 minutes before baking.

For waffles, preheat waffle iron, add batter, sprinkle with berries, and follow manufacturer's instructions for baking.

For pancakes, spoon or pour batter onto a preheated lightly oiled griddle. Sprinkle with berries and cook until top is bubbly; turn and cook until bottom side is done. Serve piping hot with selected toppings.

Makes 3 waffles or about 18 4-inch pancakes.

1 ½	cups rolled oats
¼	cup buckwheat flour
1 ¼	cup whole wheat flour
¾	cup polenta or coarse cornmeal
4	teaspoons baking powder
1 ½	teaspoons baking soda
2 ½	cups low-fat buttermilk
2	cups plain nonfat yogurt
3	tablespoons real maple syrup
2	eggs, lightly beaten, or 4 egg whites, lightly beaten
⅓	cup vegetable oil
1	cup fresh whole blueberries or sliced strawberries

Blue Corn or Polenta Pancakes

Take your choice: blue cornmeal yields dark color and distinct flavor; polenta is yellow and crunchy.

Mix cornmeal or polenta, flour, baking soda, and salt together in a medium-sized bowl.

In a small bowl, combine honey, buttermilk or yogurt, oil, and egg. Mix quickly into dry ingredients; let stand for 10 minutes. Pour onto lightly greased hot griddle, sprinkle on fruit, and cook until tops are bubbly. Turn and cook until bottoms are done. Serve hot with favorite toppings.

Serves 4.

SAVORY VARIATION: Omit fruit and add 3 to 4 tablespoons grated onion and ¼ cup freshly grated Parmesan cheese to the batter. Serve with a spicy tomato relish or Fresh Tomato Salsa (page 54).

1 ½	cups coarse blue cornmeal or polenta
¼	cup unbleached all-purpose flour
1	teaspoon baking soda
½	teaspoon salt (optional)
1	tablespoon honey
2	cups buttermilk or yogurt
2	tablespoons vegetable oil
1	egg, lightly beaten
1	cup blueberries or other fruit

Polenta and Sesame Muffins

In large mixing bowl, combine polenta or cornmeal, flour, wheat germ, sesame seeds, baking soda, and salt.

Combine eggs, buttermilk, and oil in medium-sized bowl. Pour into cornmeal mixture, stirring with fork just until moistened. Spoon into greased muffin pans or preheated corn stick pans, filling ¾ full. Bake at 425° F until golden, about 20 minutes.

Makes about 12 muffins.

1 ¼	cup polenta or coarse cornmeal
½	cup whole wheat flour
⅓	cup plain wheat germ
½	cup sesame seeds
1	teaspoon baking soda
1	teaspoon salt (optional)
2	eggs, slightly beaten
1 ¼	cup buttermilk
⅓	cup plus 1 tablespoon oil

Polenta Pancakes reward good workout.

Apricot Bran Muffins

2	cups bran
1	cup raisins
¾	cup chopped dried apricots
⅓	cup molasses
⅓	cup honey
1 ½	cups buttermilk
2	eggs, slightly beaten
½	cup vegetable oil
1	teaspoon vanilla
¾	cup unbleached all-purpose flour
⅓	cup whole wheat flour
2	teaspoons baking powder
1	teaspoon baking soda
1	teaspoon freshly grated nutmeg
2	teaspoons ground cinnamon
1	teaspoon ground allspice
1	cup chopped nuts (optional)

My friend Stephen Marcus suggests an overnight soaking of bran with spices in buttermilk to develop flavors and soften bran for a luxurious muffin. Here's a variation on his recipe.

Combine bran, raisins, apricots, molasses, honey, and buttermilk in a mixing bowl. Cover and let stand overnight.

In the morning combine eggs, oil, and vanilla in a small bowl and add to the bran mixture. Combine flours, baking powder, nutmeg, cinnamon, allspice, and nuts in a separate bowl, stir and add to the bran mixture, mixing quickly and lightly. Spoon into greased muffin tins, filling about ¾ full. Bake at 425° F until browned, about 25 minutes.

Makes about 16 muffins.

Apricot Bran Muffins are tasty anytime treats.

Ginger and Banana Bran Loaf

2½   cups puréed or mashed very ripe bananas (about 4 large bananas)

¾   cup low-fat buttermilk

1½   cups bran

2   eggs, lightly beaten

4   egg whites, lightly beaten

½   cup vegetable oil

1   tablespoon molasses

½   cup whole wheat flour

½   cup buckwheat flour

½   cup light brown sugar

1   tablespoon baking powder

1   teaspoon salt (optional)

1   tablespoon cinnamon

1   cup currants or raisins

4   teaspoons finely chopped crystallized or preserved ginger

½   cup chopped cashews or peanuts

An easy low-fat bread for breakfasts or snacks with the flavor of the tropics.

Mix bananas, buttermilk, and bran together in large bowl. Let stand 2 or 3 minutes before adding eggs, egg whites, oil, and molasses; beat well.

Combine flours, sugar, baking powder, salt, and cinnamon. Quickly stir along with raisins, ginger, and nuts into banana mixture, stirring only enough to moisten. Pour batter into lightly oiled 9- x 5-inch loaf pan and bake at 350° F until golden brown and a pick inserted in the center comes out clean, about 1 hour.

Makes 1 loaf.

Whole Wheat Pizza

I was delighted to have my own suspicions confirmed about one of my favorite foods. Pizza has been described as the near-perfect food, as it supplies basic nutrients in almost exact amounts recommended in the Dietary Goals outlined by the McGovern Committee on Nutrition and Human Needs: 15% protein, 27% fat, and 58% carbohydrate.

To simulate brick pizza ovens that turn out crisp crusts, lay unglazed quarry tiles or a special pizza stone on the bottom rack in the lowest position of the oven. Slide pizza directly onto the preheated tiles for baking using a wide spatula or special pizza peel made for transferring pizzas.

Make individual pizzas as suggested below for lunch, light dinner main course, or anytime snacks. Try tiny crusts for appetizer-sized pizzas.

To make the crust, dissolve the sugar or honey in warm water, stir in yeast and let stand until bubbling, about 5 minutes.

Combine flours in large mixing bowl. Add salt if desired. Make a well in the center, pour in yeast mixture and 3 tablespoons olive oil. Stir with a wooden spoon until flour is incorporated and dough begins to form a mass.

Turn dough onto a lightly floured surface and lightly knead by hand until dough looses its stickiness, about 5 minutes. Continue kneading until dough feels elastic and appears smooth and shiny, about 5 minutes longer. Shape into a ball and place in a lightly oiled bowl, turning to coat dough with oil. Cover with a cloth towel or plastic wrap and set to rise in a warm place until doubled in bulk, about 1 to 1½ hours.

1	tablespoon sugar or honey
¾	cup warm water
1	tablespoon active dry yeast
2	cups unbleached all-purpose flour
1	cup whole wheat pastry flour
½	teaspoon salt (optional)
¼	cup virgin olive oil, plus extra for drizzling
	Cornmeal for dusting
	Toppings (see next page)

Selected Toppings:

- Thinly sliced ripe tomatoes, skim-milk mozzarella, and whole basil leaves
- Rich tomato sauce, Parmesan, and fresh or dried oregano
- Slivered sun-dried Italian tomatoes and mild goat cheese
- Thinly sliced red onions, tomatoes, and Parmesan cheese
- Shredded cooked chicken breast, julienned leeks, fresh tarragon or sage, and a little Gruyère cheese
- Browned thin eggplant slices, roasted red pepper, Herbes de Provence, and a sprinkle of Parmesan
- Sliced fresh mushrooms, preferably wild types, and low-fat cheese
- Steamed clams or mussels, garlic, fresh herbs
- Small oysters, minced shallots, very thinly sliced lemon
- Cottage Cheese Topping (page 54), fresh caviar, and chives (put on all toppings after crust is fully baked; some immediately)
- Pesto sprinkled with pine nuts
- Cooked Mexican-seasoned beans, fresh or canned jalapeño peppers, low-fat cheese, and salsa

Again, punch down risen dough. Knead for about 30 seconds and divide into 4 to 6 equal pieces. Form each piece into a ball. At this point dough balls may be wrapped in plastic and frozen until needed. Always thaw before proceeding.

Place each ball of dough on a lightly floured surface and roll in a circle to ¼ inch thick. Fold about ½ inch of edges under, pinching dough together to form a raised wall. Place rounds on a baking sheet sprinkled with cornmeal. Cover with a lightweight towel and let rise for about 20 minutes while oven preheats to 400° F.

Lightly brush dough with olive oil. With wide spatula or pizza peel, slide onto preheated tiles or pizza stone placed on lowest oven rack and bake for 7 to 8 minutes. Remove from oven, add selected toppings, drizzle with more oil, and bake until crust is well browned and any cheese is melted, about 7 to 10 minutes more. Sprinkle with fresh herbs, if desired, and serve piping hot.

Makes 4 to 6 individual pizzas.

Pizza with fresh tomatoes, golden peppers, low-fat mozzarella, truffles, and fresh herbs.

Fresh Tomato Salsa

1 ½	cup peeled and chopped ripe tomato
½	cup finely chopped red onion
¼	cup seeded, deveined, and finely minced mild or hot chili pepper, more or less according to taste
1-2	tablespoons minced or pressed garlic
¼	cup chopped fresh cilantro

Salsa perks up grilled fish, chicken, or burgers, and adds freshness to Black Bean Chili (page 62) and other soups. Use as a dip for low-fat corn chips made by placing cut pieces of corn tortillas on a baking sheet at 375° F, turning once or twice until crisp, about 5 minutes.

Combine all ingredients. Let stand for about 1 hour to blend flavors. Use same day it is made.

Makes about 2 cups.

Cottage Cheese Topping

1	cup low-fat cottage cheese
3	tablespoons low-fat buttermilk or nonfat milk
1	tablespoon freshly squeezed lemon juice

Use this low-fat blend in traditional recipes that call for sour cream.

Place all ingredients in food processor or blender and run until smooth. Refrigerate in covered container for up to one week.

Makes 1 cup.

Quesadillas

Make these Mexican-style toasted sandwiches with low-fat cheese slices; read labels to be sure palm or coconut oil does not replace milk fat. Vary by adding bits of cooked chicken, beans, cilantro, or other ingredients between the cheese.

Moisten tortillas on both sides with cold water. Add half of the cheese slices to four tortillas, then top each with equally distributed chopped peppers and salsa. Add a dollop of Cottage Cheese Topping to each and cover with remaining cheese slices and tortillas. Toast on both sides in a preheated skillet or griddle until tortillas are lightly browned and cheese is melted. Cut into quarters to serve.

Serves 8 as appetizer; 4 as part of main course.

8	corn or flour tortillas
8	slices low-fat cheese
¼	cup chopped fresh or canned mild or hot chili peppers, more or less according to taste
4	tablespoons fresh or canned hot or mild salsa
	Cottage Cheese Topping (see recipe, opposite)

Roasted Garlic

Squeeze sweet, creamy garlic out of browned husk and spread on pieces of warm crusty whole grain bread.

Heat griddle or heavy skillet over moderately-high heat until water dances when sprinkled on surface. Place unpeeled garlic cloves on non-oiled hot surface, pressing down gently with spatula. Turn and roast until all sides are charred with splotches of brown.

Serves 2 to 4.

1	large head garlic, outer husks removed and cloves separated

Middle Eastern Garbanzo Spread

Puréed garbanzo beans mixed with sesame seed paste, or *Hummus Bi Tahini*, is a quickly prepared, complete-protein appetizer. Use as a spread or dip with pieces of whole wheat pita bread or with vegetables cooked until crisp-tender and then quickly chilled in iced water and drained. Thinned down with lemon juice and oil, *hummus* is a tasty dressing for salads of greens or vegetables. Combine with plain yogurt for a tangy sandwich spread.

Reheat garbanzo beans to boiling point, cool slightly, and place in food processor or blender with a bit of their liquid. Purée until smooth. Add garlic, tahini, reserved garbanzo liquid with enough water to equal 1 cup, ¼ cup olive oil, and lemon juice. Blend to creamy consistency, adding a little more liquid if necessary.

Place in a small serving bowl or spread thinly on a plate; drizzle with olive oil and garnish with pomegranate seeds, if available, and mint leaves. Surround with lavash crackers, pita bread cut into small wedges, or vegetables.

Makes about 4 cups

4	cups cooked garbanzo beans (see recipe, page 43), or 2 1-pound cans garbanzo beans, drained, reserve liquid
2	tablespoons minced or pressed garlic
1	cup canned tahini (toasted sesame seed paste, available in Greek or Middle Eastern specialty stores and most gourmet or health food markets)
¼	cup olive oil, plus extra for drizzling over top
½	cup freshly squeezed lemon juice
	Pomegranate seeds (garnish)
	Fresh mint leaves (garnish)
	Lavash crackers, pita bread, or crisp vegetables

Garbanzo beans and sesame seeds team up in
Hummus Bi Tahini.

Herbed Yogurt Dipping Sauce

1   cup plain nonfat yogurt

1   tablespoon minced or pressed garlic

¾   teaspoon grated horseradish

1   tablespoon freshly squeezed lemon juice

2   tablespoons minced parsley or other fresh herb

2   tablespoons minced chives

Freshly ground black pepper

Serve as a dip or sauce for tender raw vegetables, or vegetables that have been steamed until crisp-tender, quickly cooled in iced water, and drained.

Combine all ingredients in food processor or blender and purée until smooth. Cover and let stand for about 5 hours to allow flavors to blend.

Makes about 1 cup.

Yogurt Cheese

2   cups plain nonfat or low-fat yogurt

Chopped herbs (optional)

Pepper or other spices (optional)

Sugar and fresh berries (optional)

Make the cheese, then go savory by adding herbs, pepper, spices, or garlic. Or go sweet by serving with a little sugar or honey and fresh berries.

Place yogurt in a cone-type coffee filter with paper liner; cover top with plastic wrap and let drain over a bowl for about 12 hours. Whey will drip into the bowl, leaving a creamy cheese. Discard whey and use cheese plain, or add herbs and spices to taste for a savory spread. Or sprinkle with a little sugar or drizzle with honey and surround with fresh berries; spread on breakfast or tea breads.

Makes 1 cup.

Warm Bean Toast Appetizers

In warm weather, toast the bread pieces over a hot grill instead of in the oven and enjoy these nutritious tidbits while barbecuing.

Spread bread pieces on a baking sheet and place in 300 degree oven until crisp and lightly browned, about 15 minutes. Remove from oven and reserve.

Heat oil in medium-sized saucepan over medium heat and sauté garlic and rosemary until garlic is soft, about 2 minutes. Remove from heat.

Combine beans and stock in food processor and purée until smooth. Add to saucepan with garlic, add cayenne and black peppers to taste, and cook over medium heat until mixture is thickened, about 10 minutes. Remove from heat and stir in lemon juice.

To serve, spread warm bean mixture on toasted bread pieces. Add dabs of Cottage Cheese Topping and Tomato Salsa to each piece and garnish with herb. Arrange on a serving tray, and pass while warm.

Serves 8 to 10.

½	crusty whole wheat French- or Italian-style baguette, thinly sliced, cut into approximately 2-inch squares
2	tablespoons olive oil
1	teaspoon minced or pressed garlic
1	tablespoon minced fresh rosemary leaves, or 1 teaspoon crumbled dried leaves
2	cups cooked dried black, white or red beans (see recipe, page 43), drain well
¾	cup homemade chicken stock or regular-strength canned broth, heated until warm
	Cayenne pepper
	Freshly ground black pepper
2	tablespoons freshly squeezed lemon juice
	Cottage Cheese Topping (see recipe, page 54)
	Fresh Tomato Salsa (see recipes, page 54)
	Fresh rosemary, flat Italian parsley leaflets, or cilantro

SOUPS

Snow Pea Soup
with Red Pepper Purée

2	medium-sized red sweet peppers
2	tablespoons vegetable oil
6	green onions, including tops, chopped
1 ½	pounds fresh snow peas or other edible pod peas, strings removed
	Freshly ground white pepper
5	cups hearty-flavored homemade vegetable or chicken stock or regular-strength canned broth
1	cup plain nonfat yogurt or low-fat milk

Swirls of roasted red pepper create interesting patterns on top of a pale green background.

Broil or place peppers over an open flame, turning until skins are blackened all over, about 5 minutes. Place charred peppers in a paper bag, closing loosely for about 10 minutes. Rub off charred skin, cut in half, remove and discard stems and seeds, and chop peppers. Purée in blender or food processor until smooth. Strain through a medium sieve and reserve. Can be made a day ahead and reheated just before using.

Heat oil in a saucepan over medium-high heat and sauté onions and peas until somewhat soft and bright green, about 5 minutes. Add stock or broth and bring to a boil. Reduce heat to low and simmer until peas are tender, about 20 minutes. Add pepper to taste.

Place soup, in small batches, in blender or food processor and purée until smooth. Strain and rub through a coarse sieve, discarding vegetable pieces. Combine strained soup with yogurt or milk and heat to simmering.

Ladle into individual bowls and spoon in fresh pepper purée to form a swirled pattern.

Serves 4 to 6.

Swirls of red pepper top Snow Pea Soup.

Black Bean Chili

2 tablespoons oil

1 cup chopped onion

1 cup chopped celery

1 cup chopped red or green pepper

2 tablespoons minced green chili pepper

4 cups fresh tomatoes, peeled, seeded, and chopped, or 3 cups canned Italian-style tomatoes with juices

4 cups cooked black, pinto, or mixed beans (see recipe, page 43)

2-3 cups frozen tofu, thawed and crumbled (optional)

3 tablespoons chili powder

2 teaspoons ground cumin

1 tablespoon minced or pressed garlic

Cayenne pepper

Plain yogurt

Fresh Tomato Salsa (see recipe, page 54)

Once tofu has frozen, thawed, drained, and squeezed of excess moisture it assumes the texture of ground beef. I've suggested it as an optional addition to this dish if you want additional protein. Expand into a main dish by serving with Quesadillas (page 55).

Heat oil in large heavy pan over moderate heat and cook onion, celery, and fresh peppers until soft, about 15 minutes. Add tomatoes, cooked beans, tofu if desired, chili powder, cumin, garlic, and pepper to taste. Reduce heat and simmer about 25 minutes.

Serve with dollops of yogurt and Fresh Tomato Salsa.

Serves 4.

Tomato Garlic Soup

Vine-ripened tomatoes and fresh basil are essential for this creamy summer treat made with low-fat milk. Forget it if you have to use hard plastic supermarket varieties. If you can afford to add the calories and fat, it's ethereal with at least part cream.

Heat oil, margarine, or butter in saucepan over medium heat. Add garlic, reduce heat to low, and simmer until garlic is soft and sweet, about 8 minutes. Add 4 cups chopped tomatoes. Increase heat to medium-high and cook until tomatoes are heated through and release their juices, about 5 minutes.

Transfer tomato-garlic mixture to bowl, and working in batches, purée tomato-garlic mixture in food processor or blender, returning to saucepan.

Add milk and pepper to taste. Stir in basil and remaining chopped tomatoes. Chill at least two hours to develop flavors. Remove from refrigerator early enough to serve slightly chilled.

Serves 4.

2	tablespoons oil or unsalted margarine or butter
4-6	tablespoons minced or pressed garlic
6	cups chopped peeled ripe tomatoes with their juices
1	cup low-fat milk
	Freshly ground pepper
½	cup chopped fresh basil

Sushi Salad

¼   cup rice vinegar

3   tablespoons sugar

1   teaspoon salt (optional)

2   cups short-grain brown rice (cooked according to directions on page 41)

5   fresh shiitake mushrooms, sliced, or 3 dried shiitake mushrooms

¼   cup dried hijiki seaweed

2   tablespoons reduced-sodium soy sauce

½   cup canned bamboo shoots, cut into thin julienne

1   medium-size carrot, coarsely shredded

1   small cucumber, coarsely chopped

1   cup fresh green vegetables (cut asparagus or green beans, peas, or edible pod peas), blanched till crisp-tender, then quickly chilled in ice water, and drained

   Sliced pickled ginger (garnish)

   Toasted sesame seeds (see note; garnish)

Sushi Salad with Japanese tidbits is excellent picnic fare.

Although seldom found in American sushi bars, this form of presentation known as *Chirashi-Zushi,* which translates "scattered sushi," is authentically Japanese. It's almost as easy to prepare as any pasta or grain salad. Instead of enclosing sushi rice in edible wrappers or topping it with a piece of fish, vegetables are tossed with the sweet-sour rice and served on plates.

In small saucepan, heat vinegar, sugar, and salt over high heat until sugar is dissolved. Cool briefly.

Fluff warm rice once and scoop into wide shallow bowl. Pour vinegar mixture over rice, folding carefully with wooden spatula to avoid breaking rice kernels. To make rice glossy, toss continuously while fanning with a paper or straw fan or hair dryer set on cool until rice is cool and all liquid absorbed. Cover and store at room temperature; do not refrigerate.

If using dried mushrooms, soak in warm water until softened, about 25 minutes. Drain, squeezing out liquid. Discard stems and slice.

Soak hijiki seaweed in soy sauce and enough warm water to cover for about 30 minutes, drain, rinse, and drain again.

Combine mushrooms, most of the seaweed (save some for garnishing), bamboo shoots, carrot, cucumber, and green vegetable in a small bowl. Gently stir into seasoned rice, being careful not to mash rice. Place on serving platter and garnish with pickled ginger slices, reserved seaweed, and toasted sesame seeds. Serve soon after preparation; do not refrigerate.

Serves 8 to 10.

NOTE: To toast sesame seeds, place them in a small heavy skillet over moderate heat. Stir until they begin to turn golden. Remove from heat and pour onto a plate to cool.

Curried Vegetable Rye Salad

2   cups fresh broccoli florets, cut into bite-sized pieces

2   cups fresh cauliflower florets, cut into bite-sized pieces

2   cups plain nonfat yogurt

1   tablespoon minced or pressed garlic

2   tablespoons minced fresh ginger

3   tablespoons curry powder, more or less to taste

   Freshly ground black pepper

4   cups cooked whole rye berries or groats (see recipe, page 41)

6   green onions, thinly sliced

1   cup white raisins

Cooked whole millet, brown rice, or other grains can substitute for rye.

Steam broccoli and cauliflower separately until just barely tender; immerse immediately in ice water to halt cooking and retain color. Drain.

In medium-sized mixing bowl, combine yogurt, garlic, ginger, curry powder, and pepper to taste. Reserve.

In large mixing bowl, combine rye, onions, raisins, and well-drained broccoli and cauliflower. Stir in curried yogurt dressing. Serve slightly chilled or at room temperature.

Serves 8 to 10.

Bulgur and Parsley Tabbuli Salad

3   cups boiling water

½   cup bulgur (precooked cracked wheat), washed and drained

1½   cups finely minced parsley, about 2 bunches

½   cup minced green onions

3   medium-sized ripe tomatoes, peeled and finely chopped

¼   cup freshly squeezed lemon juice, more or less to taste

¼   cup olive oil

   Freshly ground black pepper

Middle Eastern *tabbuli* normally has more bulgur than this California recipe, which emphasizes a great quantity of parsley.

Pour boiling water over bulgur in a medium-sized bowl, cover, and let soak for at least 1 hour. Drain in a cheescloth-lined sieve, gathering cheesecloth into a bag and squeezing out excess liquid.

Combine well-drained bulgur with remaining ingredients in a large mixing bowl and season to taste. Cover and refrigerate for at least 4 hours for bulgur to absorb flavors.

Serves 6.

Summer Corn and Rice Salad

One hot summer day only leftover corn-on-the-cob, a little brown rice, a bunch of basil, and some fresh tomatoes were left in The Rockpile Press refrigerator. Thus this tasty salad was born.

Cook corn in a large pot of boiling water just until tender, 3 to 5 minutes. Remove from pot and plunge into cold water to halt cooking. Drain well. With a sharp knife, scrape kernels off cobs.

In a large mixing bowl, combine corn, rice, tomatoes, basil, and pepper to taste. Toss with vinaigrette or plain vinegar if you're very calorie conscious. Serve at room temperature.

Serves 4 to 6.

3	large ears of corn
1½	cups cold cooked brown rice (see recipe, page 41)
2	large ripe tomatoes, peeled and chopped
1	cup, tightly packed, small fresh basil leaves
	Freshly ground pepper
	Balsamic Vinaigrette (see recipe below)

Balsamic Vinaigrette

A sprinkling of Balsamic vinegar is my daily salad standby. When vinaigrette is called for, this less-oil version is my favorite.

In a small bowl, whisk together vinegar, water, garlic, mustard, sugar, and pepper to taste. Slowly drizzle in ¼ cup oil, whisking to combine. Stir in herb.

Makes about ¾ cup.

¼	cup Balsamic vinegar
¼	cup water
1	clove garlic, minced or pressed
1	tablespoon Dijon-style mustard
1	teaspoon light brown sugar (optional)
	Freshly ground black pepper
¼	cup olive or vegetable oil
	Chopped fresh basil, coriander (cilantro), tarragon, or other herb (optional)

Mexican Tostada Salad

This Mexican-style salad is great for lunch or light suppers. Small portions can be served as a first course of a Tex-Mex dinner.

Soak beans overnight. Place in large, heavy pot along with onion, carrots, garlic, chili powder, and pepper to taste. Bring to boil over moderately-high heat, reduce heat, and simmer until beans are tender, 3 to 6 hours, depending on type of bean. Let most of the liquid cook away. Mash beans with potato masher or place in food processor and purée briefly until chunky.

Heat tortillas at 400° F until lightly brown and crisp. Remove to plate and spread with hot beans. Add cooked chicken if desired. Top with shredded lettuce, tomatoes, radishes, green onions, green chili or tomato salsa, a dollop of Cottage Cheese Topping, and a sprinkling of Parmesan cheese.

Serves 6.

2	cups dried pinto, black, or red kidney beans, rinsed and picked over
2	cups chopped onion
1	cup finely chopped carrot
1-2	tablespoons minced or pressed garlic
3	tablespoons chili powder, more or less to taste
	Freshly ground black pepper
6	large corn tortillas
2	cups cooked shredded or chopped chicken breast (optional)
1	head iceberg lettuce, thinly shredded
2	firm ripe tomatoes, chopped
6-8	radishes, thinly sliced
5-6	green onions, tops included, thinly sliced
	Fresh Tomato Salsa (see recipe, page 54) or good quality bottled red or green Mexican-style salsa
	Cottage Cheese Topping (see recipe, page 54)
¼	cup freshly grated Parmesan cheese

All ages, including eleven-year-old Lucy, enjoy tostadas.

Curried Tuna and Buckwheat

¼ cup plain nonfat yogurt

¼ cup mayonnaise

2 teaspoons freshly squeezed lemon juice

2 tablespoons hot or mild curry powder, more or less according to taste

1 7-ounce can water-packed tuna, drained

¾ cup cooked buckwheat groats or kasha (see basic recipe, page 39)

½ cup cored and chopped tart green apple

½ cup currants or raisins

3 tablespoons chopped raw cashews

Serve this low-fat, high protein, and complex carbo mix on a bed of greens or stuffed into halved whole wheat pita bread or on lightly toasted whole grain bread. When I need a quick unplanned lunch, this is a reliable standby, omitting the cooked buckwheat if there's none in the refrigerator.

In a small mixing bowl, combine yogurt, mayonnaise, lemon juice, and curry powder. Toss with tuna, buckwheat groats, apple, currants, and cashews. Serve on lettuce leaves or as sandwich filling in whole wheat or pita bread.

Serves 3 or 4.

New Potato and Green Bean Salad

1 pound tiny new potatoes

¼ cup low-fat cottage cheese

1 tablespoon nonfat buttermilk

¼ teaspoon freshly squeezed lemon juice

1 tablespoon mayonnaise

3 tablespoons minced chives
  Freshly ground black pepper

½ pound tiny green beans, or larger beans cut into 1-inch pieces, cooked until crisp-tender, quickly chilled in iced water, and drained

½ cup chopped pistachio nuts

Start with the smallest and freshest potatoes and beans you can find.

Place potatoes in a saucepan, barely cover with water, bring to a boil, cover, and cook until almost tender, about 8 minutes. Drain and slice.

Combine cottage cheese, buttermilk, lemon juice, mayonnaise, and chives in a food processor or blender and process until creamy. Add pepper to taste. Toss with cooked potato slices and green beans. Sprinkle with pistachios. Serve at room temperature.

Serves 6 to 8.

Grilled Tuna Salad Niçoise

Although canned water-packed tuna is hard to beat nutritionally as a quick meal, the fresh counterpart is a whole new ball game. Try this variation on the Niçoise theme as a warm summer evening meal. Shark or other firm-fleshed fish are excellent alternatives.

Roast peppers over hot coals, turning frequently, until skin is blackened. Transfer to paper bag and close loosely for 15 minutes. Peel away charred skin, cut in half, remove seed, and slice into thin strips.

Brush tuna pieces with oil and cook over hot coals until just barely done on the inside, certainly no longer than 3 to 4 minutes on each side.

Arrange plates with greens, potatoes, beans, tomatoes, grilled peppers, and warm tuna. Garnish with parsley and nasturtiums. Drizzle with Balsamic Vinaigrette, or, if you're watching calories, with plain Balsamic vinegar.

Serves 4.

2	red, green, or gold sweet peppers
½	pound fresh tuna, filets or steaks, boned, skinned, and cut into 4 pieces
1	tablespoon olive or vegetable oil
	Tender young salad greens
4	small new potatoes, boiled, cooled, and sliced
½	pound tender green beans, steamed until crisp-tender, then quickly cooled in ice water, and drained
2	Italian plum tomatoes, sliced lengthwise
	Balsamic Vinaigrette (see recipe, page 67)
	Fresh parsley, preferably flat Italian type (garnish)
	Nasturtium blossoms (garnish)

Asparagus Salad
with Asian-style Peanut Dressing

This dressing or sauce goes well with well-drained cooked spinach, steamed or grilled broccoli or eggplant, or crisply blanched shredded cabbage.

In a blender or food processor, combine peanut butter, sesame oil, soy sauce, mirin or sherry, lemon juice, garlic, sugar or honey, yogurt, and chili oil to taste; purée until smooth. Stir in 1 tablespoon sesame seeds. Serve asparagus with dressing on side; sprinkle remaining sesame seeds on top.

Serves 4 to 6.

NOTE: To toast sesame seeds, place them in a small heavy skillet over medium heat, stirring frequently, until golden. Remove from heat and pour onto a plate to cool.

¼	cup freshly ground chunky peanut butter
2	teaspoons Oriental-style sesame oil
3	tablespoons reduced-sodium soy sauce
1	teaspoon mirin (Japanese sweet wine) or sherry
½	teaspoon freshly squeezed lemon juice
2	teaspoons minced or pressed garlic
1-2	tablespoons brown sugar or honey
½	cup plain low-fat yogurt
	Szechuan chili oil
4	teaspoons toasted sesame seeds (see note)
1	pound fresh asparagus, sliced diagonally into 2-inch pieces, cooked until crisp-tender, then quickly chilled in iced water, and drained

Fresh Asparagus Salad with Asian-style Peanut Dressing.

Minted White Bean Salad

1	pound cannellini or other white beans, rinsed and picked over
3	cloves garlic
1	stalk celery
1	large carrot
½	cup white wine vinegar
⅓	cup olive or vegetable oil
⅓	cup chopped fresh mint
	Freshly ground white pepper

Simple and refreshing.

Soak beans overnight in enough water to cover. Place in large pot with garlic, celery, and carrot. Bring to boil, reduce heat and simmer, adding extra water if needed, until beans are tender, about 3 hours. Drain beans. Remove and discard vegetables.

In a large bowl, whisk together vinegar, oil, mint, and pepper to taste. Add beans and toss to mix. Cover and let stand for 1 hour before serving. Best eaten the same day.

Serves 4 to 6.

Black-eyed Pea Salad

1	cup chopped onion
5	whole cloves
6	whole peppercorns
1	bay leaf
1	pound dried black-eyed peas, rinsed and picked over
4	green onions, tops included, thinly sliced
1	tablespoon Dijon-style mustard
¼	cup freshly squeezed lemon juice
⅓	cup olive oil
2	tablespoons minced or pressed garlic
1	teaspoon minced fresh oregano, or ¼ teaspoon dried
	Freshly ground black pepper
	About 1 pound fresh small dandelion, spinach, or other greens, tough stems removed, washed, and dried

Lentils or split peas can substitute for black-eyes.

Tie onion, cloves, peppercorns, and bay leaf in double layer of cheesecloth and place in a cast-iron Dutch oven or other heavy pot along with peas. Cover with cold water and bring to a boil over high heat. Reduce heat and simmer, uncovered, until peas are done but firm, about 30 minutes. Discard cheesecloth bag, drain peas, and place in large mixing bowl. When cool, toss with green onions.

In a small bowl, whisk together mustard and lemon juice. Slowly whisk in oil; stir in garlic and oregano. Pour over peas and toss until well coated. Add pepper to taste. Serve at room temperature on bed of greens. If made ahead, covered, and refrigerated, allow to return almost to room temperature before serving.

Serves 8 to 12.

Pasta Verde

Green pasta, green sauce, and green vegetables are a prelude to dinner of chicken or fish or the main vegetarian event.

Wash spinach and place in a saucepan; cook until wilted. Combine spinach and whatever liquid has formed from cooking in food processor or blender along with basil, garlic, pepper, cottage cheese, and one-half of the Parmesan cheese. Purée until fairly smooth.

Steam or blanch broccoli and peas separately until crisp-tender. Reserve.

Cook pasta in 4 quarts boiling water until *al dente.* Drain and immediately toss with spinach-basil sauce. Top with broccoli and peas; sprinkle with Parmesan cheese and garnish with whole basil leaves. Serve hot, passing additional cheese at the table if you wish.

Serves 8 as first course, 4 to 6 as main course.

2	bunches fresh spinach
1	cup tightly packed fresh basil
2	teaspoons minced or pressed garlic
	Freshly ground black pepper
⅔	cup low-fat cottage cheese
½	cup freshly grated Parmesan cheese
1	pound fresh broccoli, florets only
1	pound fresh peas, shelled, or frozen tiny peas
12	ounces spinach pasta
	Whole small fresh basil leaves

Basil Tofu Pesto

Allow to stand long enough for bland tofu to take on flavors of other ingredients.

Place tofu in a saucepan, cover with water, and boil 10 minutes; drain. Cut tofu into pieces and combine with oil, garlic, and basil in food processor or blender; process until fairly smooth. Stir in cheese. Allow to stand for at least 1 hour to develop flavor. When ready to serve, stir in a bit of hot water, and toss with hot pasta cooked al dente. Garnish with additional basil leaves if desired.

Makes about 1½ cups.

¼	pound tofu
3	tablespoons olive oil
1	tablespoon chopped garlic
2	cups tightly packed fresh basil leaves
¼	cup freshly grated Parmesan cheese
	Fresh basil leaves (garnish)

Tomato Pasta
with Avocado Sauce

Avocado should be treated as an occasional source of fat instead of a frequent fruit or vegetable. Here, in a dish adapted from Jerry Needle's original, avocado replaces oil or butter in an uncooked mélange that blends into a creamy sauce when tossed with piping hot noodles. Make only when you have flavorful tomatoes.

Mash the avocado in a mixing bowl. Toss in onion, tomato, chopped basil, and pepper to taste. Reserve.

Cook pasta in 4 quarts boiling water until *al dente*. Drain and toss in a heated bowl with reserved sauce. Garnish with basil leaves and cherry tomatoes. Pass Parmesan cheese to sprinkle over top.

Serves 3 to 4.

1	large very ripe avocado, pitted and scooped from peel
½	cup finely chopped sweet red onion
1	cup ripe, good-tasting peeled and chopped tomato
1	cup tightly packed chopped fresh basil
	Freshly ground black pepper
12	ounces fresh tomato pasta or dried whole wheat pasta
	Whole fresh basil leaves (garnish)
	Cherry tomatoes (garnish)
	Freshly grated Parmesan cheese

Fresh Tomato Sauce

Naturally sweetened with carrots, this is a basic sauce for pasta, rice or other grains, or for adding to numerous dishes. Add favorite chopped fresh herbs if you like and substitute canned Italian-style tomatoes when fresh ones aren't available.

Heat oil in saucepan over low heat and cook onion, celery, and carrot until vegetables are soft, about 25 minutes. Add garlic and tomatoes; simmer until slightly thickened, about 35 minutes. Add pepper to taste.

Serve as is or purée in a food processor or blender for a smoother sauce.

Makes about 2 cups.

¼	cup olive or vegetable oil
1	cup finely chopped onion
1	cup finely chopped celery
1	cup finely chopped carrot
2	tablespoons minced or pressed garlic
4	cups peeled and chopped ripe flavorful tomatoes
	Freshly ground black pepper or dried red pepper flakes

Hot tomato-flavored pasta is ready for tossing with avocado sauce ingredients.

Pasta with Vegetable Garden Sauce

½ pound fresh snow peas, sugar snaps, or other edible pod peas, one type or assorted

½ pound baby summer squash or zucchini, with blossoms if possible, or small squash sliced

½ pound tender fresh asparagus, cut into 2-inch pieces

1½ cups shelled fresh or frozen tiny peas

6 ounces spinach tagliarini or other thin pasta

6 ounces whole wheat tagliarini or other thin pasta

¼ cup extra virgin olive oil

1 large red or gold sweet pepper, cut into julienne

1 tablespoon minced or pressed garlic

½ cup homemade chicken stock or regular-strength canned broth

1½ cups chopped yellow tomatoes or halved yellow pear tomatoes

¼ cup minced fresh chives

1½ cups fresh basil, cut into chiffonade or chopped

½ cup freshly grated Parmesan cheese

Fresh basil sprigs (garnish)

Shown on the cover, this dish combines spinach and whole wheat pastas with garden fresh vegetables, best-quality olive oil, and sweet fresh Parmesan cheese.

Separately steam or blanch pea pods, squash, asparagus, and peas until crisp-tender. Drain and reserve.

Cook pastas separately in 2 quarts boiling water until *al dente*. Drain and keep warm.

Meanwhile, heat oil in heavy saucepan over medium heat. Add pepper and cook until limp, about 10 minutes. Reduce heat to low, add garlic, and cook about 3 minutes. Add stock or broth, tomatoes, chives, and basil chiffonade; increase heat to medium and cook until tomatoes give off juice. Add reserved vegetables and cook until hot. Toss with pasta. Sprinkle with Parmesan cheese and garnish with fresh basil sprigs. Pass additional cheese at table.

Serves 4 to 6.

Pasta Frittata

For those occasional egg-based main dishes, Italian open-faced omelets are easy to make and can be served warm or at room temperature. Cooked whole wheat pasta adds complex carbohydrates; its a great way to use up any leftover pasta, with or without sauce, but so good its worth cooking pasta just for this dish.

In a medium-sized mixing bowl, combine eggs, cheese, pasta, and pepper to taste.

Heat oil and butter in a 12-inch non-stick skillet over medium heat. Sauté onion and garlic until golden. Stir in vegetables and distribute evenly on bottom of pan. Pour egg mixture over vegetables and turn heat to lowest position. Cook until eggs are set around the edges. With a spatula, gently lift edges of the omelet and tilt pan, allowing uncooked egg to run underneath. Continue cooking until eggs are nearly set. Place skillet under a preheated broiler for about 30 seconds to set the top, making sure frittata doesn't burn.

Turn frittata onto a plate. Cut into wedges to serve.

Serves 4.

6	eggs, well beaten
¼	cup freshly grated Parmesan cheese
½	pound whole wheat pasta, cooked *al dente*
	Freshly ground black pepper
1	tablespoon olive oil
1	tablespoon unsalted butter or margarine
½	cup chopped onion
2	teaspoons minced or pressed garlic
1-2	cups steamed vegetables, cut into bite-sized pieces

Pasta with Truffled Wild Mushroom and Garbanzo Sauce

2	tablespoons olive oil
½	cup minced shallots
2	teaspoons minced or pressed garlic
8	ounces fresh chanterelles or other wild mushrooms (available in autumn), or 1½ ounces dried wild mushrooms, soaked in hot water to cover until soft, drained (reserve liquid), and chopped
4	cups peeled and chopped flavorful ripe yellow or red tomatoes
½	teaspoon or more ground saffron (optional)
	Freshly ground white pepper
4	cups cooked garbanzo beans, drained (see recipe, page 43)
1	pound fresh tagliatelle or dried thin pasta
	Fresh or preserved black or white truffles, very thinly sliced
	Freshly grated Parmesan cheese or crumbled goat cheese

Simple garbanzo beans are transformed with earthy wild mushrooms and truffles into a sauce suitable for the fanciest dinner party.

Heat oil in a large saucepan over medium heat and sauté shallots until soft, about 10 minutes. Add garlic and mushrooms (if using dried mushrooms, add soaking liquid) and cook for about 2 minutes. Add tomatoes, saffron, and pepper to taste. Reduce heat to low and simmer, uncovered, until sauce is slightly thickened, about 15 to 20 minutes.

Add garbanzo beans and cook until heated through, about 10 minutes.

Meanwhile cook pasta in 4 quarts boiling water until *al dente*. Drain, and toss immediately with sauce. Serve topped with truffle slivers according to taste and budget. Offer cheese at the table.

Serves 8 as first course; 4 to 6 as main course.

Begin a special meal with small portions of pasta tossed with garbanzo beans, yellow tomatoes, truffles, and a variety of wild mushrooms.

Seafood Lasagna

1	tablespoon vegetable or olive oil
1	large leek, thinly sliced, or 4 green onions, part of green tops included, thinly sliced
3	teaspoons minced or pressed garlic
½	cup dry white wine
1	pound fresh small scallops
½	pound lasagna noodles, preferably freshly made whole wheat
2	tablespoons cornstarch
1	cup nonfat milk
1	cup fish stock or bottled clam juice
	Freshly ground white pepper
1	cup part-skim ricotta cheese
½	cup low-fat cottage cheese or crumbled tofu
½	cup freshly grated Parmesan cheese
½	pound cooked and shelled tiny shrimp
1	pound fresh cooked crab meat, flaked

Unorthodox sauce and cheese for a lasagna, but the result is creamy rich, low-fat, and nutritionally sound.

Heat oil in skillet over medium heat and sauté leek or onions until soft, about 2 to 3 minutes. Reduce heat to low, add garlic and white wine, and simmer for 5 minutes. Add scallops and simmer until centers are opaque, about 2 to 3 minutes. Pour scallops into a sieve set over a bowl; let drain well, reserving juices and scallops separately.

Cook lasagna noodles in 4 quarts boiling water until *al dente,* about 10 to 12 minutes. Drain and rinse with cold water until cool; drain again and reserve.

To make white sauce, combine cornstarch with cold milk and stock or clam juice in a small saucepan. Place over low heat and cook, stirring, until sauce thickens. Add reserved juices from poaching scallops and pepper to taste; reserve.

In a small bowl, combine ricotta, cottage cheese or tofu, and Parmesan cheese. Set aside.

Lightly oil bottoms and sides of a 9- by 12-inch baking dish and layer ⅓ of the reserved noodles on bottom. Add ⅓ of the white sauce, spreading evenly over noodles. Top with ⅓ each of the poached scallops and their cooking vegetables, cooked shrimp, and crab meat. Add ⅓ of the cheese mixture. Repeat layering until ingredients are used up, ending with cheese on top.

Cover baking dish with foil and bake at 350° F for 20 minutes; remove foil and cook until cheese is melted and begins to color on top. Remove from oven and let stand for about 15 minutes before cutting into squares.

Serves 12 as first course, 6 as main course.

Cajun Seafood Gumbo

The gummy texture from cooking green okra pods acts as a thickening and flavoring agent in this original Louisiana hearty soup based on French bouillabaisse. Some cooks prefer thickening the soup with filé, powdered leaves of Gulf Coast native sassafras trees known to the Choctaw Indians as *"kumbo."* If okra is unavailable or not a household favorite, omit it and add 1 ½ teaspoons filé powder, available at gourmet groceries, just prior to serving. Do not allow gumbo to boil after filé has been added. If you wish, add 1 pound chopped baked ham or sliced hot smoked sausages along with stock for more authentic flavor.

In a 1-gallon stockpot, make a roux by blending ½ cup oil with flour, stirring constantly to prevent burning, over low heat until very dark brown, about 30 minutes. Add celery, onion, sweet pepper, parsley, and garlic. Cook, stirring frequently, for 45 minutes to 1 hour.

In a large skillet, cook okra in remaining oil until browned, then add to gumbo mixture. (At this stage, the mixture can be cooled and refrigerated up to 48 hours, or frozen for later use.) Add stock or broth, water, Worcestershire, catsup, tomato, bay leaf, thyme, and rosemary along with liquid red pepper, black pepper, and red pepper flakes to taste. Simmer 3 to 4 hours.

About 30 minutes before serving time, add crab and shrimp; simmer just until heated through. Stir in oysters and their liquid during the final 10 minutes of simmering. Add molasses or sugar and lemon juice. If you have not used okra, add filé at this point. Taste and correct seasonings. Serve over mounds of boiled or steamed rice in large soup bowls. Sprinkle with parsley.

Serves 6 as a main course; 10 as first course.

⅔	cup vegetable oil
½	cup unbleached all-purpose flour
2	cups chopped celery
2	cups chopped onion
1	cup chopped red or green sweet pepper
¼	cup chopped parsley
1	tablespoon minced or pressed garlic
¾	pound okra, sliced
1	quart homemade chicken or fish stock or regular-strength canned chicken broth
1	quart water
¼	cup Worcestershire sauce
¼	cup catsup
1	cup peeled and chopped ripe tomato
1	bay leaf
½	teaspoon thyme
¼	teaspoon rosemary
	Liquid red pepper seasoning
	Freshly ground black pepper
	Dried red pepper flakes
1	pound cooked crab meat, flaked
1 ½	pounds cooked shrimp, shelled and deveined
½	pound shucked oysters
1	tablespoon molasses or brown sugar
2	tablespoons lemon juice
3	cups cooked brown rice (see recipe, page 41)
	Chopped fresh parsley, preferably flat Italian type (garnish)

Salmon and Shellfish Baked in Parchment

Have everything ready so that parchment packets can go directly from oven to plate to table. You'll want good crusty whole wheat French bread to soak up every drop of the delicious juices.

Cut parchment or butcher's paper into heart-shaped pieces, using full width of paper. Rub lightly with vegetable oil or spray lightly with vegetable oil spray. Set aside.

Blanch carrot, leek, and red pepper strips in boiling water until crisp-tender, about 1 minute. Cool in ice water and drain well; dry on paper toweling. When dry distribute vegetables equally over one-half of each of the four pieces of parchment and top with a piece of fish. Equally distribute scallop slices and prawns over fish.

In a small bowl, combine wine and lime juice; pour 1 tablespoon over each piece of fish. Drizzle with oil and season to taste with pepper. Sprinkle fish with minced herb, topping with sprig of herb and lime.

Fold the other half of the heart over the fish and tightly seal packet by making a series of overlapping folds. Cook on a baking sheet in a 475° F oven until packets are puffed up, about 10 minutes. Immediately place packets on plates and serve, allowing diners to cut open their packet at the table and inhale the aromatic burst of flavors.

Serves 4.

4	20-inch squares of cooking parchment or heavy duty butcher's paper
	Vegetable oil or vegetable oil spray
1	medium-sized carrot, cut into very thin julienne
1	medium-sized leek, cut into very thin julienne, or 4 scallions, cut into very thin julienne
1	medium-sized red sweet pepper, cut into very thin julienne
4	salmon steaks or filets (3 to 4 ounces each) or other firm-fleshed fish such as halibut, sea bass, shark, or tuna
1	pound sea scallops, sliced if large
8	medium-sized prawns, shelled
2	tablespoons dry white wine
2	tablespoons freshly squeezed lime juice
4	teaspoons vegetable or olive oil
	Freshly ground black pepper
2	tablespoons minced fresh herb (basil, chervil, cilantro, parsley, or tarragon)
4	sprigs of same fresh herb
4	thin slices lime

Individual parchment packets are opened to reveal salmon with prawns, scallops, vegetables, lime, and fresh dill.

Thai-Style Grilled Oysters

2	tablespoons Oriental-style dark sesame oil
3	tablespoons vegetable oil
3	tablespoons reduced-sodium soy sauce
½	cup rice wine vinegar
2	tablespoons freshly squeezed lemon juice
	Szechuan hot chili oil
	Crushed red pepper
24	medium-sized oysters, shucked and drained

Oysters are normally grilled intact until their shells pop open. Here, a spicy vinaigrette doubles as marinade and dipping sauce for shucked oysters.

Combine sesame and vegetable oils with soy sauce, vinegar, lemon juice, and as much chili oil and red pepper as your tastebuds can stand; blend well. Pour over oysters in a medium-sized bowl and marinate, covered and refrigerated, for 4 to 6 hours, stirring or shaking several times.

Drain oysters, reserving marinade. Cover grill with a piece of wire mesh to prevent oysters from falling through slits. Cook oysters over hot coals until just done, about 1 to 2 minutes per side. Serve with small bowls of marinade for dipping sauce.

Serves 2 to 3 as main course; 4 to 6 as appetizer.

Steamed Fish with Black Bean Sauce

4	1½- to 2-inch steaks sea bass or other firm fish
4	tablespoons Oriental-style dark sesame oil
¼	cup plus 2 teaspoons reduced sodium soy sauce
8	thin slices fresh ginger root, slivered
2	green onions, cut into 1-inch pieces and slivered

Instead of fish steaks, small whole fish that will serve one or two make an attractive presentation. Products not found on your grocer's shelves are available in Oriental markets.

Wash fish, pat dry with paper towel, and place in bowl. Combine sesame oil, ¼ cup soy sauce, slivered ginger, and green onion, and pour over fish.

Rinse beans twice in warm water; drain well. Crush with minced ginger and add remaining 2 teaspoons soy sauce and sherry. Rub over fish; place in shallow bowl. Place in steamer over simmering water and cook until fish is done and flakes easily with a fork, 10 to 15 minutes, depending on thickness and type of fish.

Serves 4.

½	cup canned black beans
4	thin slices ginger root, minced
1	tablespoon sherry (optional)

Glazed Garlic Chicken and Tofu

Don't be startled by two heads of garlic; the finished result is deliciously sweet. If you're partial to one or the other, the dish can be prepared by doubling either chicken or tofu instead of the combination. In any case, serve over cooked Japanese buckwheat noodles, any thin whole wheat pasta, or any cooked whole grain, and watch garlic lovers swoon.

Combine garlic, pepper flakes, soy sauce, honey, and vinegar in a shallow bowl. Add tofu and set aside for about 20 minutes.

Heat oil in a large skillet over medium-high heat and sauté chicken breasts, turning to brown on all sides, about 3 minutes total. Remove chicken with tongs or slotted spoon and set aside. Remove and reserve tofu, pouring garlic mixture into skillet. Cook, stirring frequently, until the sauce has thickened and partly reduced, about 10 to 15 minutes. Add chicken and/or tofu to sauce and cook, stirring constantly, until pieces are slightly glazed, 1 or 2 minutes.

Spoon chicken, tofu, and sauce over cooked noodles. Sprinkle with sesame seeds and pass cilantro at the table.

Serves 4 to 6.

2	heads fresh garlic, broken into cloves, peeled, and coarsely chopped
2	teaspoons dried red pepper flakes, more or less according to taste
½	cup reduced-sodium soy sauce
5	tablespoons honey
1½	cups rice wine vinegar
12	ounces firm tofu, cut into long ½-inch wide strips
3	tablespoons vegetable oil
2	whole chicken breasts, halved, boned, skinned, and cut lengthwise into ½-inch strips
½	pound buckwheat noodles, cooked *al dente*
¼	cup toasted sesame seeds
	Chopped fresh cilantro

Whole Wheat Loaf
with Basil Chicken Stuffing

3　tablespoons olive oil

1　chicken, fat and skin removed, cut up

1　cup homemade chicken stock or regular-strength canned chicken broth

1　1-pound loaf crusty whole wheat French- or Italian-style bread (round or oblong)

1　teaspoon minced or pressed garlic

2　tablespoons fresh parsley, preferably flat Italian-type

⅔　cup tightly packed fresh basil leaves

¼　cup toasted almonds or pine nuts (see note)

1　egg, lightly beaten

¼　cup freshly squeezed lemon juice

　　Freshly ground black pepper

Great piping hot from the oven or still slightly warm. Wrap in foil and take along for nutritious picnicking.

Heat olive oil in large skillet over medium heat and sauté chicken pieces until browned on all sides. Add half of the stock or broth, cover, and cook over low heat until the meat almost falls off the bone, about 50 minutes. Add more broth during cooking if necessary. Remove chicken pieces, cool, discard skin and bones, and shred or cut into bite-sized pieces; reserve liquid.

Half bread horizontally and pull out most of the soft inside, reserving both crust pieces and insides.

In food processor, combine garlic, parsley, basil, and nuts; process until well chopped and blended. Add soft bread insides and ½ cup liquid from chicken skillet and purée.

Combine puréed mixture with egg, lemon juice, reserved chicken, and pepper to taste. Spoon into bottom bread crust and cover with top crust. Bake at 350° F until chicken mixture is heated through and bread is crisp, about 20 minutes. Serve hot or at room temperature, cutting into wedges or thick slices.

Serves 4 to 6.

NOTE: To toast pine nuts, place them in a small heavy skillet over moderate heat. Stir until they begin to turn golden. Remove from heat and pour onto a plate to cool.

Wedge of basil- and chicken-stuffed bread goes anywhere.

Chicken with Wild Rice

1	tablespoon olive oil
1	tablespoon unsalted butter or margarine
3	chicken breasts, halved, boned, and skinned
6	green onions, thinly sliced
½	cup red, gold, or green sweet pepper
½	cup chopped celery
½	pound mushrooms, preferably shiitake, sliced
2	teaspoons minced or pressed garlic
1	cup uncooked wild rice
2	cups homemade chicken stock or regular-strength canned chicken broth
1	tablespoon chili powder
2	tablespoons freshly squeezed lemon juice
	Freshly ground black pepper
	Liquid red pepper seasoning

Easy and delicious.

Heat oil and butter or margarine in medium-sized skillet over medium-high heat. Sauté chicken until lightly browned, about 2 minutes on each side. Remove chicken with slotted spoon and set aside. To the skillet, add onions, peppers, and celery, reduce heat to medium, and cook just until wilted. Add mushrooms, garlic, and rice, stirring until coated. Stir in chicken stock or broth. Cover and simmer, without stirring, for 25 minutes.

Remove cover and add chili powder, lemon juice, and black pepper and liquid pepper to taste; stir. Place chicken breasts on top of rice mixture. Cover and bake at 350° F for 1 hour.

Serves 6.

Grilled Yogurt-crusted Cornish Hens with Couscous

Removing skins from already lean cornish hens cuts the fat in half. Yogurt marinade tenderizes and forms a great crust during grilling. In lieu of Cornish hens, chicken pieces may be used. Serve with *couscous,* **tiny Middle-eastern style wheat pasta.**

In blender or food processor, combine yogurt, garlic, ginger, and oil. Purée until smooth, then blend in curry powder to taste. Pour over hens and marinate, covered, in refrigerator for at least 12 or up to 24 hours. Occasionally stir or shake container to coat thoroughly.

Remove from refrigerator about 30 minutes before cooking. Grill over hot coals until browned and done, about 10 minutes on each side. Serve hot or at room temperature with couscous and condiments.

Serves 4 to 8.

2	cups plain nonfat yogurt
2	tablespoons minced or pressed garlic
2	tablespoons chopped ginger root
2	tablespoons olive oil
¼	cup curry powder, mild or hot, more or less, according to taste
4	Cornish hens, split in half and skin removed
	Couscous, cooked according to package directions
	Condiments: plain yogurt, chopped cilantro, chutney, chopped peanuts, toasted sesame seeds

Corn Pudding in Spicy Turkey Crust

Based on a classic South American dish, this lighter variation relies on ground turkey instead of more traditional beef. If you enjoy red meat, feel free to substitute your choice. Sweet and creamy, freshly picked corn is ideal, but frozen kernels are acceptable.

Heat oil in large skillet over medium-high heat, add onion, celery, sweet pepper, chili pepper, and tomato. Stirring often, cook until juices have evaporated and vegetables are soft, about 10 minutes. Add garlic, raisins, cumin, chili powder, and pepper flakes to taste; stir and remove from heat.

Combine ground turkey and bread crumbs and add to vegetable mixture, stirring until well mixed. Spoon into 10-inch pie plate or heavy skillet that can go into the oven, patting to cover bottom and extending up sides like a pie shell.

Combine corn, flour, eggs, milk, and green onions and pour into meat shell, spreading evenly. Bake at 350° F until center feels set when touched, about 35 minutes. Remove from oven and let stand about 15 minutes before serving. Garnish with cilantro and cut into wedges.

Serves 6 to 8.

1	tablespoon vegetable oil
1	cup chopped onion
½	cup chopped celery
½	cup chopped red or green sweet pepper
2	tablespoons minced chili pepper
1½	cups peeled, seeded, and chopped tomato
1	tablespoon minced or pressed garlic
½	cup raisins
1	teaspoon ground cumin
1	tablespoon chili powder
	Red pepper flakes
1	pound ground turkey
¼	cup fine whole wheat bread crumbs
2¾	cups corn, cut from cob (about 4 ears)
1	tablespoon unbleached, all-purpose flour
2	eggs, beaten
½	cup nonfat milk
2	green onions, including tops, thinly sliced
	Fresh cilantro sprigs

Creamy fresh corn pudding fills a spicy crust made of ground turkey.

Turkey Breast
with Winter Squash and Fruits

½	cup vegetable or olive oil
½	cup red wine or fruit-flavored vinegar
1	teaspoon Dijon-style mustard
2	tablespoons minced or pressed garlic
2	tablespoons minced fresh or preserved ginger
1	tablespoon dried thyme
2	teaspoons dried rosemary
2	teaspoons ground cumin
	Freshly ground white pepper
2	tablespoons water-packed green peppercorns
¾	cup coarsely chopped dried figs or dates
¾	cup dried pitted prunes
½	cup dried apricots, quartered
½	cup dried pears, sliced
3	pounds boneless turkey breast, cut into 2-inch pieces
3	cups bite-sized pieces peeled butternut or other winter squash
¾	cup homemade chicken stock or regular-strength canned broth
½	cup dry white wine
1	tablespoon grated orange rind
1	tablespoon grated lemon rind
¾	cup toasted broken hazelnuts (see note)

Superlean turkey breasts, available whole or in pieces, are marinated and baked with dried fruits and fresh squash. Try this very flavorful concoction for a change-of-pace Thanksgiving or other cool weather feast.

Combine oil, vinegar, mustard, garlic, ginger, thyme, rosemary, cumin, pepper to taste, and peppercorns in a mixing bowl. Add dried fruits and turkey pieces. Cover and refrigerate at least 6 hours, preferably overnight.

Pour turkey mixture into large shallow casserole or baking pan, stir in squash pieces, add stock and wine, cover with tight-fitting lid or foil, and bake at 350° F for 20 minutes.

Remove cover and bake until turkey and squash are tender, about 25 minutes. With a slotted spoon, remove turkey, squash, and fruit to a heated platter. Place pan on high heat and cook until sauce is thickened and slightly reduced. Drizzle sauce over turkey. Sprinkle with orange and lemon rind and nuts. Serve with wild rice or other whole grain tossed with sautéed fresh mushrooms.

Serves 6 to 8.

NOTE: To toast hazelnuts, place them in a small skillet in a 350° F oven, stirring frequently, until lightly browned, about 8 to 10 minutes. Remove from oven and pour onto a plate to cool; rub between fingers to remove skins.

Venetian-style Calves' Liver

Studies reveal that people who exercise strenuously need extra iron. Liver is the best source of iron but is so high in cholesterol that it should be eaten only very occasionally. Liver also stores toxins, so you should look for liver from very young calves. This is the best way I know to prepare tender calves' liver combined with another power food, onions, cooked slowly to bring out their natural sweetness.

3   tablespoons olive oil

3   cups thinly sliced yellow onions

1½   pounds calves' liver, sliced ¼-inch thick

Freshly ground black pepper

Small fresh sage leaves (garnish)

Heat oil in a large skillet over medium heat. Add onions and cook, stirring frequently, until limp. Reduce heat to very low and continue to cook, stirring occasionally, until onions are golden and almost caramelized, about 35 minutes. Remove onions from skillet with a slotted spoon, leaving oil in pan. Onions can be cooked ahead and reheated.

Trim any gristle and skin tissue from liver slices and cut into thin, bite-sized strips. Heat oil in skillet over high heat; when quite hot, add liver pieces and cook until just barely browned. Turn strips, add onions, and pepper to taste. Stir mixture together and cook for no more than a minute. Transfer to a warm serving dish, sprinkle with sage leaves, and serve while piping hot, preferably with polenta, or over broad homemade tomato or whole wheat noodles.

Serves 4.

Szechuan Fried Rice with Vegetables

A perpetual favorite quick-and-easy one-dish supper of mine that always satisfies. Feel free to vary the vegetables, using whatever is on hand. Add cooked chicken or shellfish if you wish.

Steam broccoli, zucchini, and carrots separately until crisp-tender. Reserve.

In a wok or large heavy skillet, heat vegetable and sesame oils over medium heat, add onion, reduce heat to low and cook until onion is soft. Add sweet pepper and continue cooking until onion is lightly golden. Add garlic, soy sauce, hot chili oil to taste (please be quite generous), reserved vegetables, and rice. Increase heat to medium-high and stir-fry until vegetables and rice are heated through, about 3 to 4 minutes. Serve sprinkled with sesame seeds and a little cilantro. Pass additional chili oil and cilantro at the table.

Serves 4 as main course; 6 to 8 as side dish.

NOTE: To toast sesame seeds, place them in a small heavy skillet over moderate heat. Stir until seeds are golden. Remove from heat and pour onto a plate to cool.

½	pound broccoli, florets separated and stems sliced
1	medium-sized zucchini, diagonally sliced
2	medium-sized carrots, diagonally sliced
2	tablespoons vegetable oil
	2 tablespoon Oriental-style sesame oil
1	small onion, halved and sliced
1	small red or green sweet pepper, cut into thin strips
1	tablespoon minced or pressed garlic
⅓	cup reduced-sodium soy sauce
	Szechuan hot chili oil
2	cups cooked brown rice (see directions on page 41)
¼	cup toasted sesame seeds (see note)
	Chopped fresh cilantro

Szechuan Fried Rice is a quick and flavorful complete meal.

Bean-filled Tamale Pie

¼	cup vegetable oil
1½	cups chopped onion
¾	cup chopped celery
1	cup chopped red or green sweet pepper
1½	tablespoons minced or pressed garlic
3	tablespoons chili powder, more or less according to taste
2	teaspoons ground cumin
5	cups peeled, seeded, and chopped fresh tomatoes or canned Italian-style tomatoes, with their juices
3	cups whole corn kernels, preferably freshly cut from cob
¾	cup tomato paste
3	cups cooked red, pinto, or black beans (see recipe, page 43)
	Cayenne pepper
5	cups homemade chicken stock or regular-strength canned broth or water
2½	cups coarse yellow cornmeal
½	cup freshly grated Parmesan cheese
1	cup grated part-skim mozzarella or Jack cheese
	Paprika
	Chopped fresh cilantro

Grain and beans team once again to form a delicious complete vegetable protein. A pound of ground turkey or very lean beef can be sautéed along with onions instead of or in addition to the beans.

Heat oil in a large skillet over medium heat; add onion, celery, and sweet pepper. Cook until vegetables are soft, about 20 minutes; add garlic, and cook briefly. Stir in chili powder, cumin, tomatoes, corn, and tomato paste. Cook, stirring frequently, for about 20 minutes. Stir in cooked beans and cayenne pepper to taste.

Bring stock, broth, or water to a boil in a large saucepan. Slowly pour in cornmeal, stirring constantly to prevent lumps. Continue cooking and stirring until thick, about 10 minutes. Cool until it can be touched with hands.

Dampen hands with water and line the bottom and sides of a lightly oiled ovenproof casserole with about two-thirds of the cornmeal mixture. Spoon in the bean filling. Press remaining cornmeal mixture with damp hands and cover top of filling, pressing to seal sides. Sprinkle with cheeses and paprika and bake at 350° F for about 1 hour. Serve hot; pass cilantro to sprinkle on top.

Serves 8.

Creole Red Beans and Rice

Traditionally laden with ham, this meatless version offers considerably less calories. If you aren't counting or want even more protein than the balance that beans with rice provides, throw in a ham bone and a big slab of baked ham when you add the beans to the pot. In New Orleans, white rice is always served; after eating it with better-tasting brown rice, I can't go home again.

Soak beans overnight in enough cold water to cover.

Heat oil in cast-iron Dutch oven or heavy pot. Add onion, celery, and sweet pepper, and cook until soft. Add beans with their soaking liquid, green onions, garlic, stock or broth, parsley, bay leaves, thyme, oregano, cumin, and peppers to taste. Bring to boil over high heat, then turn heat to low and simmer until beans are tender and start breaking up, about 3 hours. Add water during cooking if necessary.

Stir in tomato paste and Worcestershire sauce. Taste and correct seasonings. Remove about 1 cup of the beans and purée them in the food processor or mash with fork; return to pot. (This helps thicken the gravy, which is normally thickened with marrow from ham bones). Serve over mounds of hot cooked brown rice. Garnish with parsley.

Serves 6.

1	pound dried red kidney beans, rinsed and picked over
2	tablespoons vegetable oil
2	cups chopped onion
2	cups chopped celery
1	cup chopped green sweet pepper
3	green onions, including tops, thinly sliced
1-2	tablespoons minced or pressed garlic
2	cups homemade chicken or beef stock or regular-strength canned broth
3	tablespoons minced fresh parsley
2	bay leaves, broken into pieces
2	teaspoons dried thyme
1	teaspoon dried oregano
1	teaspoon ground cumin
	Cayenne pepper
	Dried red pepper flakes
	Freshly ground black pepper
2	tablespoons tomato paste
2	tablespoons Worcestershire sauce
4	cups cooked brown rice (see recipe, page 41)
	Fresh parsley, preferably flat Italian type (garnish)

Creole Three-onion Pie

⅛   cup whole wheat pastry flour

⅛   cup unbleached all-purpose flour

½   teaspoon baking powder

⅛   teaspoon salt (optional)

2   teaspoons unsalted butter or margarine

2   tablespoons nonfat milk

3   tablespoons vegetable oil

1   cup chopped yellow or white onion

½   cup thinly sliced leek

½   cup chopped green onion

1   egg, lightly beaten

4   egg whites, lightly beaten

1   cup Cottage Cheese Topping (see recipe, page 54)

     Freshly ground black pepper

     Cayenne pepper

½   cup sunflower seeds

     Whole chives (garnish)

Serve with seafood or poultry. Bacon usually tops the pie, but I've used sunflower seeds, which toast up brown and crunchy like bacon minus the fat. Be sure to spoon up any liquid left in the pie dish after serving and pour over wedges.

To make the biscuit crust, sift flours, baking powder, and salt together into a mixing bowl. Add butter or margarine and milk, mixing with a wooden spoon until well blended. Turn out onto a floured board and knead by hand for about 5 minutes. With a floured rolling pin, roll out dough to about ¼-inch thickness. Line a 9-inch pie pan, trimming to leave about ½-inch overhang, and fold edges under, pinching the overhang to form a fluted edge. Bake in 350° F oven for 5 minutes. Remove from oven and set aside.

Heat oil in heavy skillet over low heat, add onion, and cook, stirring frequently, until soft, but not browned, about 15 minutes. Add leek and green onion; cook an additional 5 minutes. Remove from heat and pour into mixing bowl. Add the egg, egg whites, Cottage Cheese Topping, and black pepper to taste; mix thoroughly, but gently.

Pour onion mixture into partially baked crust. Sprinkle with cayenne pepper to taste and sunflower seeds. Bake at 400° F until edge of crust is lightly browned, about 15 to 20 minutes. Garnish top with whole chives. With a serrated knife, cut into wedges and serve hot from the oven.

Serves 6.

Green Onions Parmesana

Excellent alongside grilled or broiled chicken breast or fish. Split leeks are sweeter alternatives.

Steam the onions over boiling water until just tender, about 8 minutes. Transfer to oven-proof dish. Top with margarine, turning to coat all onions as margarine melts. Sprinkle with pepper to taste, bread crumbs, and Parmesan cheese. Place under broiler until cheese melts.

Serves 4.

12-16	green onions, trimmed, leaving several inches of tops
2	teaspoons unsalted butter or margarine
	Freshly ground black pepper
½	cup fine whole wheat bread crumbs
⅓	cup freshly grated Parmesan cheese

Corn-stuffed Onions

Vary the idea with your favorite stuffing.

In a large saucepan, cover onions with water and boil until onions are just barely tender, but still hold their shape. Drain, cool, and cut off top quarter of each onion. Trim bottoms of onions to stand upright. Scoop out centers to create a cavity about 1½ to 2 inches in diameter. Save cut-off tops and scooped centers for another purpose.

In a mixing bowl, combine corn kernels, chives, egg whites, buttermilk, olive oil, and pepper to taste. Fill onion cavities with corn mixture, sprinkle with paprika and Parmesan cheese. Place in lightly oiled oven-proof dish and bake at 350° F until corn custard is set, about 45 minutes.

Serves 6.

6	large onions
1	cup fresh corn kernels (2 to 3 ears)
2	tablespoons minced chives
4	egg whites, lightly beaten
1	tablespoon nonfat buttermilk
1	teaspoon olive oil
	Freshly ground black pepper
	Paprika
¼	cup freshly grated Parmesan cheese

Vegetable Custard

1   pound steamed vegetables, well-drained

1   cup nonfat milk

1   egg, lightly beaten

3   egg whites, lightly beaten

2   tablespoons unbleached all-purpose flour

2   tablespoons freshly grated Parmesan cheese

½   teaspoon freshly grated nutmeg

Freshly ground black pepper

Here's a change of pace that can dress up a dinner plate. Use one kind of vegetable such as asparagus, carrots, or cauliflower for distinct color. Or mix vegetables to create differing shades such as all greens or all yellows.

Place tender cooked vegetables in food processor or blender with a little of the milk and purée until smooth. Combine with remaining milk, whole egg, egg whites, flour, cheese, nutmeg, and pepper to taste.

Distribute mixture among 8 lightly oiled ramekins or custard cups. Bake at 375° F until set, about 25 minutes. Serve in baking dishes or invert onto plate. Garnish with fresh herbs or small pieces of fresh vegetables.

Makes 6 to 8 servings.

Custards of asparagus, carrots, and parsnips were baked in individual molds.

Skillet-baked Pepper Corn

3 tablespoons corn oil

5 cups fresh corn kernels, about 7 large ears, cut from cob and juice from cobs scraped into bowl

2 tablespoons minced red or green hot chili peppers, more or less according to taste

   Freshly ground pepper

⅓ cup unbleached all-purpose flour

A Southern tradition to which I've added a bit of zing.

Pour oil in a medium-sized cast-iron or other ovenproof skillet and heat in a 425° F oven until piping hot, about 25 minutes.

Combine corn, chili peppers, black pepper to taste, and flour in a mixing bowl. Pour into hot oiled skillet, return to oven, and bake until corn forms a bottom crust, about 30 minutes. Place a serving plate over skillet and invert to remove corn. Serve hot; cut into wedges at the table.

Serves 6.

Deviled Potatoes

½ pound tiny new potatoes

2 tablespoons vegetable or olive oil

1 tablespoon unbleached all-purpose flour

1 teaspoon Dijon-style mustard

¼ cup white wine vinegar

   Freshly ground black pepper

   Cayenne pepper

   Minced chives (garnish)

   Pink pepperberries (dried fruit of *Schinus molle* tree; garnish)

A change of pace from ubiquitous baked potatoes.

Boil potatoes until just barely tender; cool and slice. Combine oil, flour, mustard, vinegar, and peppers to taste in a large skillet over medium heat. Add potatoes and gently stir until potatoes are heated through. Just before serving, sprinkle with chives and pink pepperberries.

Serves 4.

Eggplant Parmesana

Most versions of this Italian classic absorb too much olive oil. Unorthodox brushing with oil and skipping of browning stage cuts down fat. Tofu and cheese add complete protein.

Brush eggplant slices with oil and place on flat pan. Broil each side until tender, 5 to 10 minutes per side. Arrange half the slices in a flat baking dish lightly coated with vegetable cooking spray. Cover with one-half the tofu and one-half the tomato sauce.

Arrange remaining eggplant over tomato sauce. Cover with remaining tofu and tomato sauce. Sprinkle with Parmesan cheese and bake, uncovered, at 350° F until heated through and slightly browned on top, about 25 minutes.

Serves 6 to 8 as side dish; 4 as main course.

1	large eggplant, cut into ½ -inch thick slices
2	tablespoons olive oil
2	cups firm tofu, crumbled
3	cups Tomato Sauce (see recipe, page 77)
¼	cup freshly grated Parmesan cheese

Lemon Zucchini with Red Peppers

Toasted oats add crunchy nutrition.

Place oats in a heavy skillet over medium heat, stirring until they are lightly toasted, about 5 to 10 minutes. Pour onto plate to cool.

Steam zucchini and red peppers until crisp-tender. Stir together with lemon rind and black pepper to taste. Place in shallow 2-quart oven-proof dish.

Distribute toasted oats over vegetables in casserole and drizzle with oil. Bake at 400° F until heated through, about 10 minutes.

Makes 6 servings.

1	cup rolled oats
4	cups, about 2 pounds, zucchini, cut in ½ -inch thick diagonal slices
3	small red sweet peppers, cut into rings
2	tablespoons grated lemon rind (no white part)
	Freshly ground black pepper
2	tablespoons vegetable or olive oil

Multi-grain Party Pilaf

This dish was originated for a Cotton Compound party at Lake Tahoe. Like many good party dishes, it can be made ahead and heated at the last minutes.

Separately cook brown rice, wild rice, and millet; set aside. Soak bulgur, drain, and set aside.

Heat oil in a large skillet over medium heat, add celery, onions, sweet pepper, and carrot. Sauté until soft, about 15 minutes. Add mushrooms and garlic and cook until mushrooms are soft, about 6 minutes.

In a large mixing bowl, combine sautéed vegetables with reserved grains. Add parsley, pepper to taste, almonds, and sunflower seeds. Sprinkle with sesame seeds. Pour into shallow baking dish, cover with foil, and heat at 350° F until heated through, about 20 minutes.

Serves 16 to 20.

NOTE: To toast nuts and seeds, place them separately in a small heavy skillet over moderate heat. Stir until lightly browned. Remove from heat and pour onto plates to cool.

4   cups raw brown rice, cooked in stock (page 41)

1   cup raw wild rice, cooked in stock (page 41)

1   cup raw millet, cooked by toasted method (page 40)

1   cup bulgur covered with 3 cups boiling water and soaked for 2 hours, drained

½   cup vegetable or olive oil

2   cups finely chopped celery

6   green onions, part of tops included, thinly sliced

2   cups finely chopped red or gold sweet peppers

2   cups coarsely chopped carrot

¾   pound mushrooms, finely chopped

2   tablespoons minced or pressed garlic

⅔   cup minced parsley

     Freshly ground black pepper

¾   cup chopped toasted almonds (see note)

1   cup toasted sunflower seeds (see note)

½   cup toasted sesame seeds (see note)

Try flavorful mixed grains for a change-of-pace party dish.

Artichokes with New Potatoes and Rice

5	medium-sized artichokes
	Juice of 1 lemon
1	cup chopped onion
1	tablespoon minced or pressed garlic
½	cup chopped fresh parsley, preferably flat Italian type
4	cups thinly sliced small new potatoes
	Freshly ground black pepper
1½	cup mixed brown and wild rice
	About 5 cups homemade vegetable or chicken stock or regular-strength canned broth, heated just to a boil
½	cup slivered oil-packed, sun-dried Italian plum tomatoes
⅓	cup whole wheat bread crumbs
½	cup freshly grated Parmesan cheese
	Cayenne pepper or sweet paprika
2	tablespoons olive or vegetable oil
	Fresh basil leaves, cut into chiffonade (optional; garnish)

Rooted in southern Italy and Spain, this earthy dish is hearty fare along with a green salad, or a nutritious accompaniment to roast chicken or grilled fish.

Cut off and discard about 2 inches from the top of each artichoke. Slice artichokes lengthwise into quarters, discarding the tough outer leaves. Cut off ends of stems and peel remaining stem with small knife or vegetable peeler. With a small knife, scrape out and discard fuzzy chokes. Slice quarters in half lengthwise and place in pot with lemon juice and enough water to cover. Simmer until just barely tender, about 12 minutes. Drain well.

Combine onion, garlic, and parsley in a small bowl and set aside.

Cover the bottom of a large lightly oiled casserole with potato slices seasoned with pepper to taste. Sprinkle with part of the onion mixture. Add a layer of rice, then artichoke hearts. Continue layering until vegetables and rice are all used, ending with potatoes on top. Add enough hot stock or broth (water can be substituted for all or part) to barely cover. Distribute tomatoes over top and sprinkle with bread crumbs, cheese, and cayenne pepper to taste. Drizzle with oil. Bake at 350° F until potatoes and rice are tender and crumb topping is golden brown, about 1 hour. Garnish with fresh basil.

Serves 8 to 10 as side dish; 4 to 6 main dish.

Fresh ingredients ready for baking into a tasty casserole.

Apple-filled Bread Pudding with Blueberry Maple Sauce

2	eggs
¼	cup brown sugar
¾	cup frozen unsweetened apple juice concentrate, thawed
2	cups low-fat milk
1	cup evaporated skim milk
5	tablespoons unsalted butter or margarine
2	teaspoons vanilla extract
1	teaspoon almond extract
½	cup white raisins
	Freshly ground nutmeg
6	ounces stale whole wheat French or other firm-textured bread, sliced ½-inch thick
3	McIntosh or other tart apples, peeled, cored, and thinly sliced
2	cups fresh or frozen blueberries
¼	cup maple sugar or syrup
1	tablespoon water
1	tablespoon freshly squeezed lemon juice

I know of no more comforting sweet to occasionally celebrate winning a race or a successful workout at the gym than all-American bread pudding. This version is full of apples with apple concentrate as part of the sweetener. It's good even without the sauce or with a little ice cream.

Combine eggs, sugar, apple concentrate, milks, 3 tablespoons butter, extracts, raisins, and nutmeg to taste. Whisk to blend well. Pour over bread slices in a large bowl and let stand, turning bread as necessary, until bread is soft and saturated, about 20 minutes.

Heat remaining 2 tablespoons butter in a large, heavy skillet over moderately high heat and sauté apple slices, stirring occasionally, until softened, about 15 minutes.

Spread about half of the bread in a lightly oiled 6- to 8-quart baking dish, cover with apple slices, and top with remaining bread; pour any remaining liquid over top. Place baking dish in a large pan and add enough warm water to reach halfway up the sides of the dish. Bake at 350° F until custard is set, about 45 to 50 minutes.

To make sauce, combine blueberries, sugar or syrup, water, and lemon juice in a small saucepan over medium heat, mashing some of the berries. Cook until blueberries are soft, about 10 minutes.

Serve warm squares of pudding with blueberry sauce or warm maple syrup. Add a tiny scoop of vanilla ice cream if desired.

Serves 6 to 8.

Fig and Date Squares

Here's a great occasional snack when you need extra energy and a piece of fruit just won't satisfy your sweet tooth.

To make filling, combine figs, dates, and lemon and orange juices in a heavy medium-sized saucepan, adding enough water to cover fruit by 1 inch. Cook over moderate heat until fruits are tender, about 20 minutes. Drain and transfer fruit to food processor. Coarsely purée fruit with lemon rind.

Meanwhile, prepare dough by combining oil, honey, and brown sugar in a mixing bowl and beat until creamy smooth. Add eggs one at a time, beating well after each addition; stir in vanilla.

In a separate bowl, combine flour, oats, baking powder, salt, and cinnamon. Add to liquid mixture and blend well.

To assemble, press and flatten one-half of the dough into a lightly oiled and floured 10¾- by 7-inch baking pan. Spread fruit filling evenly over dough. Patting lightly between slightly dampened hands, flatten and smooth remaining dough and cover filling. Bake at 350° F until lightly browned, about 25 to 30 minutes. Cool completely before cutting into squares or bars.

Makes about 24 small squares or 16 bars.

1½	cups chopped dried figs
¾	cup chopped pitted dates
2	tablespoons freshly squeezed lemon juice
¼	cup freshly squeezed orange juice
1	tablespoon grated lemon or orange rind
¾	cup vegetable oil
½	cup honey
¾	cup brown sugar
2	eggs
2	teaspoons vanilla extract
2	cups whole wheat pastry flour
2	cups rolled oats
1	teaspoon baking powder
¾	teaspoon salt (optional)
2	teaspoons ground cinnamon

New "Mystery" Pie

I grew up, as many of you probably did, with a deliciously crunchy "mystery" pie richly laced with chopped pecans and crumbled Ritz crackers. I've reduced the fat and sugar, and replaced the pecans and crackers with my reduced-fat granola. No one will believe the non-fat topping isn't made with more highly caloric cream cheese.

Combine egg whites and 1 teaspoon vanilla in a non-plastic bowl and beat until they begin to hold shape. Gradually add 1 cup sugar and baking powder, continuing to beat until stiff.

Whirl granola in food processor or blender to crumble. Fold into egg white mixture. Pour into 9-inch pie plate and bake at 350° F until set and lightly browned, about 30 minutes. Cool.

In a small bowl, using a fork or small whisk, blend yogurt cheese with remaining 2 tablespoons sugar, ginger with syrup, and remaining 2 teaspoons vanilla until creamy smooth. Spread over cooled pie and top with fruit slices. Melt marmalade in a small saucepan and brush over fruit to lightly coat. Serve garnished with mint leaves and violets.

Serves 8 to 10.

4	egg whites
1	tablespoon vanilla extract
1	cup plus 2 tablespoons sugar
½	teaspoon baking powder
2	cups Low-fat Granola (see recipe, page 44)
1	cup Yogurt Cheese (see recipe, page 58)
2	tablespoons finely chopped preserved ginger, plus ginger syrup to taste
¼	cup lime marmalade
1	cup sliced fresh figs, kiwi, or other fruit
	Mint leaves and fresh violets (garnish)

Kiwi and crisp Japanese persimmons top "Mystery" Pie made with granola and topped with gingered yogurt cheese spread.

Vermont Maple Custard

4   eggs

3   cups low-fat milk

¾   cup real maple syrup

       Pinch of salt (optional)

¼   teaspoon vanilla extract

Maple Mary McCoy shared this ultra-satisfying custard that can be enjoyed warm or cold, for breakfast, lunch, or dinner, with fresh fruit, or plain.

Beat eggs slightly, then beat in milk, syrup, salt, and vanilla. Pour into 8 small custard cups or 1-quart baking dish. Place cups or dish in larger pan and pour in 1 inch hot, not boiling, water. Bake at 300° F until a knife tests clean when inserted in the center, about 1 hour. Serve warm, at room temperature, or slightly chilled.

Serves 4 as breakfast, or 6 to 8 as dessert.

Fruit Ices

Refreshing after a long workout or competition, these slushy fruit purées are also light endings to a great meal. The variations here are suggestions to get you started on your own concoctions. All are based on fruit purées that can be frozen in ice cream makers or the freezer unit. Double the treat by serving fruit ices along with fresh fruits.

Italian ices, known as *granites,* sweetened to taste with simple sugar syrup, can be made ahead and refrigerated until needed. Of course, the type of fruit and degree of ripeness determines the amount of syrup added; some may not need any sweetener. Remember that freezing diminishes sweetness, so add a bit more syrup than you may think necessary when fruits are at room temperature.

To turn ices into lighter-textured French *sorbets,* add stiffly beaten egg whites to the cooled syrup and fruit purée. Use instead of simple syrup in accompanying recipes.

In some cases, dairy products are suggested to give a smoother texture.

If you use an ice cream freezer, pour fruit mixture into freezer container and follow manufacturer's directions. When frozen, spoon ice into plastic freezer containers, leaving at least ½- inch expansion space. Cover and place in freezer to mellow flavors, preferably overnight, but at least 5 hours. Remove from freezer about 20 minutes before serving.

In lieu of an ice cream freezer, pour fruit mixture into metal pans and place in freezer until mixture becomes firm just around the edges, about 1½ to 2 hours. Remove pan and stir frozen edges into slushy center. Return to freezer until almost frozen, about 1 hour. Spoon into food processor and process until smooth, or place in large chilled mixing bowl and beat well with electric mixer or wire whisk. Pour soft mixture into plastic freezer containers, leaving at least ½-inch expansion space. Cover and freeze for at least 5 hours.

SIMPLE SYRUP: Combine sugar and water in a small saucepan and bring to boil over medium-heat, stirring constantly until sugar is dissolved. Let cool to room temperature before pouring into storage container. Refrigerate for up to 2 weeks, using as needed. Makes 3 cups.

2	cups sugar
2	cups water

BLUEBERRY-YOGURT ICE: Place all ingredients in food processor in batches and purée until smooth, pouring into large sieve over mixing bowl. Press berries through sieve with wooden spoon, discarding skins and seed. Freeze according to directions. Makes about 5 cups.

2	pints fresh blueberries
2	cups plain low-fat yogurt
¾	cup sugar
1	tablespoon grated lemon peel

6	cups cut-up fresh ripe pineapple, shell and core removed
2	tablespoons minced drained ginger in syrup
¼	cup syrup from ginger in syrup
1 ½	cups Simple Syrup (see recipe, page 115)

GINGERED PINEAPPLE ICE: Place pineapple in batches in food processor and purée until smooth, pouring into large mixing bowl. Add remaining ingredients and freeze according to directions. Makes about 4 cups.

4	cups peeled and chopped ripe kiwis
1 ½	to 2 cups Simple Syrup (see recipe, page 115)

KIWI GRANITE: Briefly purée kiwi in food processor or blender; overprocessing brings out bitterness in seed. Combine kiwi purée and syrup and freeze according to directions. Makes about 4 cups.

2	cups freshly squeezed lemon juice
2	tablespoons grated lemon rind
3	cups Simple Syrup (see recipe, page 115)
4	egg whites, stiffly beaten

LEMON SORBET: Combine lemon juice and lemon peel in a large bowl. In a medium bowl, fold egg whites into cool syrup. Fold into lemon mixture and freeze as directed. Makes about 6 cups.

4 ½	cups seeded watermelon flesh
4	cups fresh or frozen raspberries
½ - ¾	cup Simple Syrup (see recipe, page 115)

WATERMELON-RASPBERRY GRANITE: In large mixing bowl, combine watermelon and raspberries. Place in batches in food processor and purée until smooth. Add Simple Syrup and freeze as directed.

Fresh raspberries and mint garnish Watermelon-Raspberry Granite.

Index

Recipe Index